MTTC Physics
19 Teacher Certification Exam

By: Sharon Wynne, M.S
Southern Connecticut State University

XAMonline, INC.
Boston

Copyright © 2007 XAMonline, Inc.
All rights reserved. No part of the material protected by this copyright notice may be reproduced or utilized in any form or by any means, electronic or mechanical, including photocopying, recording or by any information storage and retrievable system, without written permission from the copyright holder.

To obtain permission(s) to use the material from this work for any purpose including workshops or seminars, please submit a written request to:

XAMonline, Inc.
21 Orient Ave.
Melrose, MA 02176
Toll Free 1-800-509-4128
Email: info@xamonline.com
Web www.xamonline.com
Fax: 1-781-662-9268

Library of Congress Cataloging-in-Publication Data

Wynne, Sharon A.
    Physics 19: Teacher Certification / Sharon A. Wynne. -2nd ed.
    ISBN 978-1-58197-665-6
    1. Physics 19.   2. Study Guides.   3. MTTC
    4. Teachers' Certification & Licensure.   5. Careers

Disclaimer:
The opinions expressed in this publication are the sole works of XAMonline and were created independently from the National Education Association, Educational Testing Service, or any State Department of Education, National Evaluation Systems or other testing affiliates.

Between the time of publication and printing, state specific standards as well as testing formats and website information may change that is not included in part or in whole within this product. Sample test questions are developed by XAMonline and reflect similar content as on real tests; however, they are not former tests. XAMonline assembles content that aligns with state standards but makes no claims nor guarantees teacher candidates a passing score. Numerical scores are determined by testing companies such as NES or ETS and then are compared with individual state standards. A passing score varies from state to state.

Printed in the United States of America   œ-1

MTTC: Physics 19
ISBN: 978-1-58197-665-6

TEACHER CERTIFICATION STUDY GUIDE

Table of Contents

SUBAREA I. SCIENTIFIC INQUIRY

COMPETENCY 1.0 UNDERSTAND THE PRINCIPLES AND PROCEDURES OF SCIENTIFIC INQUIRY 1

Skill 1.1　Formulating research questions and investigations in physics 1

Skill 1.2　Developing valid experimental designs for collecting and analyzing data and testing hypotheses 1

Skill 1.3　Recognizing the need for controlled experiments 2

Skill 1.4　Understanding procedures for collecting and interpreting data to maintain objectivity 3

Skill 1.5　Recognizing independent and dependent variables and analyzing the role of each in experimental design 4

Skill 1.6　Identifying an appropriate method for presenting data for a given purpose 4

Skill 1.7　Applying mathematics to investigations in physics and the analysis of data 5

Skill 1.8　Interpreting results presented in different formats 9

Skill 1.9　Evaluating the validity of conclusions 10

Skill 1.10　Assessing the reliability of sources of information 11

COMPETENCY 2.0 APPLY KNOWLEDGE OF METHODS AND EQUIPMENT USED IN SCIENTIFIC INVESTIGATIONS. 13

Skill 2.1　Includes selecting and using appropriate measurement devices and methods for collecting data 13

Skill 2.2　Evaluating the accuracy and precision of measurement and methods collecting data 13

Skill 2.3　Identifying procedures and sources of information related to the safe use, storage, and disposal of materials and equipment related to physics investigations measurement 15

PHYSICS

Skill 2.4	Identifying hazards associated with laboratory practices and materials 16
Skill 2.5	Applying procedures for preventing accidents and dealing with emergencies 18

COMPETENCY 3.0 UNDERSTAND THE DEVELOPMENT OF SCIENTIFIC THOUGHT AND INQUIRY 20

Skill 3.1	Includes demonstrating knowledge of the reliance of scientific investigations on empirical data, verifiable evidence, and logical reasoning 20
Skill 3.2	Recognizing the effect of researcher bias on scientific investigations and the interpretation of data 21
Skill 3.3	Demonstrating an awareness of the contributions made to physics by individuals of diverse backgrounds and from different time periods 21
Skill 3.4	Recognizing the dynamic nature of scientific knowledge, including ways in which scientific knowledge changes 22

COMPETENCY 4.0 UNDERSTAND THE RELATIONSHIPS OF PHYSICS TO TECHNOLOGICAL AND SOCIETAL ISSUES, BOTH CONTEMPORARY AND HISTORICAL 24

Skill 4.1	Includes recognizing the relationships between science and technology 24
Skill 4.2	Analyzing political and social factors that influence developments in physics, including current issues and controversies related to physics research and technology 25
Skill 4.3	Evaluating the credibility of scientific claims made various forums 26

COMPETENCY 5.0 UNDERSTAND INTERRELATIONSHIPS AMONG THE PHYSICAL, LIFE, AND EARTH/SPACE SCIENCES 27

Skill 5.1	Includes recognizing major unifying themes and concepts that are common to the various scientific disciplines 27
Skill 5.2	Describing the integration and interdependence of the sciences 28
Skill 5.3	Interdisciplinary connections among the sciences and their applications in real-world contexts 29

TEACHER CERTIFICATION STUDY GUIDE

SUBAREA II.   MECHANICS

COMPETENCY 6.0   ANALYZE MOTION IN ONE AND TWO DIMENSIONS ... 30

Skill 6.1   Analyzing information related to displacement, speed, velocity, and acceleration presented in one or more representations 30

Skill 6.2   Solving problems involving constant acceleration 33

Skill 6.3   Applying principles of trigonometry and properties of vectors to analyze motion in two dimensions 34

Skill 6.4   Applying calculus to analyze motion in one dimension 39

COMPETENCY 7.0   UNDERSTAND NEWTON'S LAWS OF MOTION AND THE LAW OF UNIVERSAL GRAVITATION 43

Skill 7.1   Applying Newton's laws of motion, both descriptively and mathematically, in a variety of situations 43

Skill 7.2   Solving a variety of problems involving different types of forces in one and two dimensions 45

Skill 7.3   Analyzing the vector nature of force 49

Skill 7.4   Determining methods for measuring force and differentiating between mass and weight 50

Skill 7.5   Applying the law of universal gravitation and Kepler's laws in a variety of situations 52

COMPETENCY 8.0   UNDERSTAND CONSERVATION OF ENERGY AND CONSERVATION OF MOMENTUM 55

Skill 8.1   Includes applying the concepts of work, energy, and power in a variety of situations 55

Skill 8.2   Analyzing the kinetic and potential energy of various systems 57

Skill 8.3   Applying the principles of conservation of energy and conservation of linear momentum to situations, including elastic and inelastic collisions 59

TEACHER CERTIFICATION STUDY GUIDE

COMPETENCY 9.0 UNDERSTAND TORQUE, STATIC EQUILIBRIUM, AND ROTATIONAL DYNAMICS 62

Skill 9.1 Analyzing the forces and torques acting in a given situation 62

Skill 9.2 Applying the concepts of force, torque, and energy to analyze the operation of simple devices 64

Skill 9.3 Applying the conservation of angular momentum 65

Skill 9.4 Analyzing the motion of a rigid body in terms of moment of inertia, rotational kinetic energy, and angular momentum 66

COMPETENCY 10.0 UNDERSTAND THE CHARACTERISTICS OF OSCILLATORY MOTION 67

Skill 10.1 Analyzing models of simple harmonic motion 67

Skill 10.2 Recognizing the relationship between the simple harmonic oscillator and uniform circular motion 69

Skill 10.3 Applying the law of conservation of energy to oscillating systems 70

Skill 10.4 Recognizing the effects of damping 71

TEACHER CERTIFICATION STUDY GUIDE

SUBAREA III.   ELECTRICITY AND MAGNETISM

COMPETENCY 11.0   UNDERSTAND ELECTRIC CHARGE, ELECTRIC FIELDS, AND ELECTRIC POTENTIAL 72

Skill 11.1   Includes describing the nature of charge 72

Skill 11.2   Describing static charges in conductors and insulators 74

Skill 11.3   Applying Coulomb's law to determine forces and fields due to various charge distributions 75

Skill 11.4   Applying the concepts of electrostatic potential energy, potential, and capacitance 78

COMPETENCY 12.0   UNDERSTAND SIMPLE CIRCUITS 82

Skill 12.1   Includes describing the properties of conductors, insulators, semiconductors, and Superconductors 82

Skill 12.2   Applying Ohm's and Kirchhoff's laws to the analysis of series and parallel circuits 84

Skill 12.3   Properly using voltmeters and ammeters 87

Skill 12.4   Determining power dissipated by circuit elements 88

Skill 12.5   Analyzing energy transfer and conservation in electrical circuits 88

COMPETENCY 13.0   UNDERSTAND MAGNETIC FIELDS 89

Skill 13.1   Includes describing the properties of permanent magnets 89

Skill 13.2   Applying laws to determine the orientation and strength of a magnetic field 90

Skill 13.3   Determining the effect of a magnetic field on moving charges 92

Skill 13.4   Explaining the role of magnetic force and torque in the operation of technological devices 93

TEACHER CERTIFICATION STUDY GUIDE

COMPETENCY 14.0 UNDERSTAND ELECTROMAGNETIC INDUCTION 95

Skill 14.1 Includes finding the rate of change of magnetic flux through a surface 95

Skill 14.2 Analyzing factors that affect the magnitude of an induced emf 95

Skill 14.3 Determining the direction of an induced current or emf 96

Skill 14.4 Recognizing that magnetic energy is stored in an inductor 98

Skill 14.5 Describing alternators and the basic properties of alternating current 98

Skill 14.6 Using the principle of electromagnetic induction to explain the operation of technological devices 100

SUBAREA IV.   WAVES, ACOUSTICS, AND OPTICS

COMPETENCY 15.0 UNDERSTAND THE CHARACTERISTICS OF WAVES AND WAVE MOTION 102

Skill 15.1 Includes describing the transfer of momentum and energy by wave motion 102

Skill 15.2 Comparing longitudinal and transverse waves 103

Skill 15.3 Analyzing and relating the characteristics of waves 104

Skill 15.4 Explaining reflection, refraction, diffraction, and the Doppler effect ... 104

Skill 15.5 Applying the principle of superposition to investigate the properties of constructive and destructive interference 106

COMPETENCY 16.0 UNDERSTAND THE PRINCIPLES OF SOUND AND ACOUSTICS 107

Skill 16.1 Includes explaining the production and propagation of sound waves 107

Skill 16.2 Applying the principles of standing waves to explain resonance and to analyze the production of musical sounds in vibrating strings and air columns 107

Skill 16.3 Analyzing the relationship between sound and human perception of sound 110

PHYSICS

TEACHER CERTIFICATION STUDY GUIDE

Skill 16.4 Describing and applying the Doppler effect 110

COMPETENCY 17.0 UNDERSTAND ELECTROMAGNETIC WAVES AND THE ELECTROMAGNETIC SPECTRUM 112

Skill 17.1 Includes identifying the connection between Maxwell's equations and the generation and propagation of electromagnetic waves 112

Skill 17.2 Demonstrating knowledge of radiometry and photometry 113

Skill 17.3 Describing the electromagnetic spectrum in terms of wavelength, frequency, and energy..114

Skill17.4 Describing how the wave theory of light is applied to a variety of phenomena 115

Skill 17.5 Analyzing applications of double-slit interference, diffraction gratings, and interferometers...116

COMPETENCY 18.0 UNDERSTAND RAY OPTICS 117

Skill 18.1 Includes applying the laws of reflection, total internal reflection, and refraction 117

Skill 18.2 Using ray diagrams with lenses and mirrors......................................118

Skill 18.3 Applying the thin lens and spherical mirror equations 122

Skill 18.4 Explaining the operation of optical instruments 125

Skill 18.5 Describing the effect of limit resolution 127

PHYSICS

TEACHER CERTIFICATION STUDY GUIDE

SUBAREA V.	NATURE OF MATTER, THERMODYNAMICS, AND MODERN PHYSICS

COMPETENCY 19.0 UNDERSTAND THE PARTICULATE NATURE OF MATTER 129

Skill 19.1   Includes recognizing basic characteristics of the states of matter 129

Skill 19.2   Describing how the Maxwell-Boltzmann theory applies to an ideal gas 132

Skill 19.3   Analyzing phase changes 135

Skill 19.4   Describing the properties of materials at low temperatures 137

COMPETENCY 20.0 UNDERSTAND THE LAWS OF THERMODYNAMICS 139

Skill 20.1   Includes differentiating between temperature, internal energy, and heat 139

Skill 20.2   Calculating heat loss or gain using specific heat 139

Skill 20.3   Identifying processes of thermal energy transfer 140

Skill 20.4   Applying the principles of enthalpy, internal energy, and thermodynamic work 141

Skill 20.5   Applying the law of conservation of energy 142

Skill 20.6   Analyzing the relationship between entropy and the availability of energy to perform work 143

TEACHER CERTIFICATION STUDY GUIDE

COMPETENCY 21.0 UNDERSTAND THE BASIC IDEAS OF QUANTUM MECHANICS AND RELATIVITY 145

Skill 21.1 Includes explaining blackbody radiation and the photoelectric effect 145

Skill 21.2 Describing evidence of the dual nature of light and matter 147

Skill 21.3 Demonstrating a basic understanding of wave functions and the Schrödinger equation 148

Skill 21.4 Recognizing models of atomic structure and their relationship to spectroscopy 148

Skill 21.5 Describing the operation of lasers 152

Skill 21.6 Demonstrating a basic understanding of the theory of special relativity 154

COMPETENCY 22.0 UNDERSTAND THE BASIC IDEAS OF NUCLEAR PHYSICS 156

Skill 22.1 Recognizing models of the nucleus 156

Skill 22.2 Describing properties of nuclei and their applications 157

Skill 22.3 Relating nuclear structure and forces to radioactivity 158

Skill 22.4 Solving problems involving half-life 160

Skill 22.5 Differentiating between fission and fusion reactions and their applications 161

Skill 22.6 Calculating energy transformations in nuclear reactions 164

Skill 22.7 Demonstrating a basic understanding of the properties of quarks and the standard model of elementary particle physics 165

Sample Test 167

Answer Key 182

Rationales with Sample Questions 183

PHYSICS

TEACHER CERTIFICATION STUDY GUIDE

Great Study and Testing Tips!

What to study in order to prepare for the subject assessments is the focus of this study guide but equally important is *how* you study.

You can increase your chances of truly mastering the information by taking some simple, but effective steps.

Study Tips:

1. Some foods aid the learning process. Foods such as milk, nuts, seeds, rice, and oats help your study efforts by releasing natural memory enhancers called CCKs (*cholecystokinin*) composed of *tryptopha*n, *choline*, and *phenylalanine*. All of these chemicals enhance the neurotransmitters associated with memory. Before studying, try a light, protein-rich meal of eggs, turkey, and fish. All of these foods release the memory enhancing chemicals. The better the connections, the more you comprehend.

Likewise, before you take a test, stick to a light snack of energy boosting and relaxing foods. A glass of milk, a piece of fruit, or some peanuts all release various memory-boosting chemicals and help you to relax and focus on the subject at hand.

2. Learn to take great notes. A by-product of our modern culture is that we have grown accustomed to getting our information in short doses (i.e. TV news sound bites or USA Today style newspaper articles.)

Consequently, we've subconsciously trained ourselves to assimilate information better in neat little packages. If your notes are scrawled all over the paper, it fragments the flow of the information. Strive for clarity. Newspapers use a standard format to achieve clarity. Your notes can be much clearer through use of proper formatting. A very effective format is called the *"Cornell Method."*

Take a sheet of loose-leaf lined notebook paper and draw a line all the way down the paper about 1-2" from the left-hand edge.

Draw another line across the width of the paper about 1-2" up from the bottom. Repeat this process on the reverse side of the page.

Look at the highly effective result. You have ample room for notes, a left hand margin for special emphasis items or inserting supplementary data from the textbook, a large area at the bottom for a brief summary, and a little rectangular space for just about anything you want.

TEACHER CERTIFICATION STUDY GUIDE

3. Get the concept then the details. Too often we focus on the details and don't gather an understanding of the concept. However, if you simply memorize only dates, places, or names, you may well miss the whole point of the subject. A key way to understand things is to put them in your own words. If you are working from a textbook, automatically summarize each paragraph in your mind. If you are outlining text, don't simply copy the author's words.

Rephrase them in your own words. You remember your own thoughts and words much better than someone else's, and subconsciously tend to associate the important details to the core concepts.

4. Ask Why? Pull apart written material paragraph by paragraph and don't forget the captions under the illustrations.

Example: If the heading is "Stream Erosion", flip it around to read "Why do streams erode?" Then answer the questions.

If you train your mind to think in a series of questions and answers, not only will you learn more, but it also helps to lessen the test anxiety because you are used to answering questions.

5. Read for reinforcement and future needs. Even if you only have 10 minutes, put your notes or a book in your hand. Your mind is similar to a computer; you have to input data in order to have it processed. *By reading, you are creating the neural connections for future retrieval.* The more times you read something, the more you reinforce the learning of ideas.

Even if you don't fully understand something on the first pass, *your mind stores much of the material for later recall.*

6. Relax to learn so go into exile. Our bodies respond to an inner clock called biorhythms. Burning the midnight oil works well for some people, but not everyone.

If possible, set aside a particular place to study that is free of distractions. Shut off the television, cell phone, pager and exile your friends and family during your study period.

If you really are bothered by silence, try background music. Light classical music at a low volume has been shown to aid in concentration over other types.

Music that evokes pleasant emotions without lyrics are highly suggested. Try just about anything by Mozart. It relaxes you.

PHYSICS

7. Use arrows not highlighters. At best, it's difficult to read a page full of yellow, pink, blue, and green streaks.

Try staring at a neon sign for a while and you'll soon see my point, the horde of colors obscure the message.

8. Budget your study time. Although you shouldn't ignore any of the material, *allocate your available study time in the same ratio that topics may appear on the test.*

TEACHER CERTIFICATION STUDY GUIDE

SUBAREA I. SCIENTIFIC INQUIRY

COMPETENCY 1.0 UNDERSTAND THE PRINCIPLES AND PROCEDURES OF SCIENTIFIC INQUIRY.

Skill 1.1 Formulating research questions and investigations in physics

Although many discoveries happen by chance, the standard thought process of a scientist begins with forming a question to research. The more limited and clearly defined the question, the easier it is to set up an experiment to answer it. Scientific questions result from observations of events in nature or events observed in the laboratory. An observation is not just a look at what happens. It also includes measurements and careful records of the event. Records could include photos, drawings, or written descriptions. The observations and data collection lead to a question. In physics, observations almost always deal with the behavior of matter. Having arrived at a question, a scientist usually researches the scientific literature to see what is known about the question. Maybe the question has already been answered. The scientist then may want to test the answer found in the literature. Or, maybe the research will lead to a new question.

Sometimes the same observations are made over and over again and are always the same. For example, you can observe that daylight lasts longer in summer than in winter. This observation never varies. Such observations are called laws of nature. One of the most important scientific laws was discovered in the late 1700s. Chemists observed that no mass was ever lost or gained in chemical reactions. This law became known as the law of conservation of mass. Explaining this law was a major topic of scientific research in the early 19th century.

Skill 1.2 Developing valid experimental designs for collecting and analyzing data and testing hypotheses

Hypothesis
Once a scientific question is formulated, taking an educated guess about the answer to the problem or question is known as developing your hypothesis. A hypothesis is a statement of a possible answer to the question. It is a tentative explanation for a set of facts and can be tested by experiments. Although hypotheses are usually based on observations, they may also be based on a sudden idea or intuition. In most experiments, scientists collect quantitative data, which is data that can be measured with instruments. They also collect qualitative data, descriptive information from observations other than measurements. Interpreting data and analyzing observations are important. If data is not organized in a logical manner, wrong conclusions can be drawn. Also, other scientists may not be able to follow your work or repeat your results.

Conclusion

Finally, a scientist must draw conclusions from the experiment. A conclusion must address the hypothesis on which the experiment was based. The conclusion states whether or not the data supports the hypothesis. If it does not, the conclusion should state what the experiment *did* show. If the hypothesis is not supported, the scientist uses the observations from the experiment to make a new or revised hypothesis. Then, new experiments are planned.

Theory

When a hypothesis survives many experimental tests to determine its validity, the hypothesis may evolve into a theory. A theory explains a body of facts and laws that are based on the facts. A theory also reliably predicts the outcome of related events in nature. For example, the law of conservation of matter and many other experimental observations led to a theory proposed early in the 19th century. This theory explained the conservation law by proposing that all matter is made up of atoms which are never created or destroyed in chemical reactions, only rearranged. This atomic theory also successfully predicted the behavior of matter in chemical reactions that had not been studied at the time. As a result, the atomic theory has stood for 200 years with only small modifications.

A theory also serves as a scientific model. A model can be a physical model made of wood or plastic, a computer program that simulates events in nature, or simply a mental picture of an idea. A model illustrates a theory and explains nature. For instance, in your science class you may develop a mental (and maybe a physical) model of the atom and its behavior. Outside of science, the word theory is often used to describe someone's unproven notion about something.

In science, theory means much more. It is a thoroughly tested explanation of things and events observed in nature. A theory can never be proven true, but it can be proven untrue. All it takes to prove a theory untrue is to show an exception to the theory. The test of the hypothesis may be observations of phenomena or a model may be built to examine its behavior under certain circumstances.

Skill 1.3 Recognizing the need for controlled experiments

The design of experiments includes the planning of all steps of the information gathering activity. It is best to start any experiment with a clearly stated and understood set of goals and objectives. This will help lead the experimenter into defining the specific data that needs to be collected, how the data will be collected, and how the data will be analyzed after collection. Specifically, the experimenter should determine the number of observations needed, and over what period. The variables affecting the data collection should also be outlined and a determination of which ones will be held constant, which ones will be varied, and which ones may be out of the experimenter's control.

A scientific control improves the integrity of an experiment by isolating each variable in order to draw conclusions about the effect of the single variable in the result. As much as possible the experiments should be identical, except for the one variable being tested. If there are variables that are beyond the control of the experimenter, it is wise to identify them up front and attempt to mitigate their effect. Controls are generally one of two types, negative and positive. A negative control is used when a negative result is expected in an experiment. The negative control helps correlate a positive result with the variable being tested. For example, testing a drug on a group of rats and maintaining another group who are only given a placebo. A positive control is a sample that is known to produce a positive result to make sure the experiment is working as expected. For example printing a page with he printers own drivers before testing the printer with another program.

Once the data collection process has been defined, collection of information can begin. It is wise to check the data periodically to ensure that the data is reasonable and is being collected appropriately. However, care must be taken that data that does not necessarily fit with expectations is not simply discarded. If the data does not fit expectations, corrections may be needed to the collection method or the expected results may not be accurate.

After collection, experimental data must be analyzed, interpreted and presented in a way that can be understood by others. Data analysis may include statistical methods, curve fitting, and dividing data into subsets. Data analysis transforms the information collected during experimentation with the goal of extracting useful information and drawing conclusions. Data interpretation is the method by which the data, in its raw and analyzed forms is reviewed for meaning and explanation. It is often necessary to look at historical information in the same area of study when interpreting the data from an experiment. Presenting the data is the final step in experimental design. In this way, information is related to others interested or affected by the results of the experiment. Graphical representations are useful to communicate information, although clear and concise language is always necessary to ensure the thorough understanding of your audience.

Skill 1.4 Understanding procedures for collecting and interpreting data to maintain objectivity

Scientific data can never be error-free. We can, however, gain useful information from our data by understanding what the sources of error are, how large they are and how they affect our results. Some errors are intrinsic to the measuring instrument, others are operator errors. Errors may be random (in any direction) or systematic (biasing the data in a particular way). In any measurement that is made, data must be quoted along with an estimate of the error in it.

Precision is a measure of how similar repeated measurements from a given device or technique are. Note that this is distinguished from accuracy which refers to how close to "correct" a measuring device or technique is. Thus, accuracy can be tested by measuring a known quantity (a standard) and determining how close the value provided by the measuring device is. To determine precision, however, we must make multiple measurements of the same sample. The precision of an instrument is typically given in terms of its standard error or standard deviation. Precision is typically divided into reproducibility and repeatability. These concepts are subtly different and are defined as follows:

Repeatability: Variation observed in measurements made over a short period of time while trying to keep all conditions the same (including using the same instrument, the same environmental conditions, and the same operator)

Reproducibility: Variation observed in measurements taken over a long time period in a variety of different settings (different places and environments, using different instruments and operators)

Skill 1.5 Recognizing independent and dependent variables and analyzing the role of each in experimental design

An experiment tests the hypothesis to determine whether it may be a correct answer to the question or a solution to the problem. Some experiments may test the effect of one thing on another under controlled conditions. Such experiments have two variables. The experimenter controls one variable, called the independent variable. The other variable, the dependent variable, is the change caused by changing the independent variable. For example, suppose a researcher wanted to test the effect of vitamin A on the ability of rats to see in dim light. The independent variable would be the dose of Vitamin A added to the rats' diet. The dependent variable would be the intensity of light that causes the rats to react. All other factors, such as time, temperature, age, water given to the rats, the other nutrients given to the rats, and similar factors, are held constant.

Skill 1.6 Identifying an appropriate method for presenting data for a given purpose

In scientific investigations, it is often necessary to gather and analyze large data sets. We may need to manage data taken over long periods of time and under various different conditions. Appropriately organizing this data is necessary to identify trends and present the information to others. The uses and advantages of various graphic representations are discussed below.

Tables are excellent for organizing data as it is being recorded and for storing data that needs to be analyzed. In fact, almost all experimental data is initially organized into a table, such as in a lab notebook. Often, it is then entered into tables within spreadsheets for further processing. Tables can be used for presenting data to others if the data set is fairly small or has been summarized (for example, presenting average values). However, for larger data sets, tabular presentation may be overwhelming. Finally, tables are not particularly useful for recognizing trends in data or for making them apparent to others.

Charts and graphs are the best way to demonstrate trends or differences between groups. They are also useful for summarizing data and presenting it. In most types of graphs, it is also simple to indicate uncertainty of experimental data using error bars. Many types of charts and graphs are available to meet different needs. Three of the most common are scatterplots, bar charts, and pie charts. An example of each is shown.

Diagrams are not typically used to present the specifics of data. However, they are very good for demonstrating phenomena qualitatively.

Skill 1.7 Applying mathematics to investigations in physics and the analysis of data

Deriving and Solving Equations
Mathematics is a very broad field, encompassing various specific disciplines including calculus, trigonometry, algebra, geometry, complex analysis and other areas. Many physical phenomena can be modeled mathematically using functions to relate a specific parameter or state of the system to one or more other parameters. For example, the net force on a charged object is a function of the direction and magnitude of the forces acting upon it, such as electrical attraction or repulsion, tension from a string or spring, gravity and friction. When a phenomenon has been modeled mathematically as a set of expressions or equations, the general relationships of numbers can be applied to glean further information about the system for the purposes of greater understanding or prediction of future behavior.

The process of treating a system mathematically can involve use of empirical relationships (such as Ohm's law) or more *a priori* relationships (such as the Schrödinger equation). Given or empirically derived equations can be manipulated or combined, depending on the specific situation, to isolate a specific parameter as a function of other parameters (solution of the equation). This process can also lead to a different equation that presents a new or simplified relationship among specific parameters (derivation of an equation).

Both derivation and solution of equations can be pursued in either an exact or approximate manner. In some cases, equations are intractable and certain assumptions must be made to facilitate finding a solution. These assumptions are typically drawn from generalizations concerning the behavior of a particular system and will result in certain restrictions on the validity and applicability of a solution. An exact solution, presumably, has none of these limitations.

Another variation of approximate derivation and solution of an equation involves numerical techniques. Ideally, an analytical approach that employs no approximations is the best alternative; such an approach yields the broadest and most useful results. Nevertheless, the intractability of an equation can lead to the need for either an approximate or numerical solution. One particular example is from the field of electromagnetics, where determination of the scattering of radiation from a finite circular cylinder is either extremely difficult or impossible to perform analytically. Thus, either numerical or approximate approaches are required. The range of numerical techniques available depends largely on the type of problem involved. The approach for numerically solving simple algebraic equations, for example, is different from the approach for numerically solving complex integral equations.

Care must be taken with analytical (and numerical or approximate) approaches to physical situations as it is possible to produce mathematically valid solutions that are physically unacceptable. For instance, the equations that describe the wave patterns produced by a disturbance of the surface of water, upon solution, may yield an expression that includes both incoming and outgoing waves. Nevertheless, in some situations, there is no physical source of incoming waves; thus, the analytical solution must be tempered by the physical constraints of the problem. In this case, one of the solutions, although mathematically valid, must be rejected as unphysical.

Modeling through Algebra and Geometry

Geometry and algebra are key tools in modeling physical situations in a number of specific areas. Understanding physical systems involves examination of certain parameters of the system as some other parameter or set of parameters, such as time or a spatial dimension, is varied. Through measurements, these parameters can be quantified and plotted graphically. In many cases, the graph of the results of such an analysis will reveal a curve that relates the state of the system, in terms of some parameter, to the value of another parameter. Such a curve could be the position of a moving car with respect to time. The empirical curve can then be fitted to an algebraic expression that expresses a mathematical relationship between the parameters by way of a function. Thus, for the example of the traveling car, the position x of the car can be expressed as a function of time, x(t). For a car moving at constant speed v, the function is of the form $x(t) = v\,t + x_0$, where x_0 is an initial position at time t = 0. A similar approach can be used to model more complicated situations. Alternatively, other algebraic equations can be combined with each other, or with measured results, to derive new expressions. For empirically-based equations, curve fitting of various forms (such as least squares regression) can be employed, depending on the complexity of the data.

Algebra for physical situations may involve several different ranges or types of numbers. In some cases, the equations may involve real numbers (such as in time-domain analysis in classical electrodynamics), or they may involve complex numbers (such as in frequency-domain analysis in classical electrodynamics). Furthermore, the algebra may be discrete or continuous. Although there is some overlap, classical mechanics is based on the use of a continuous numerical scheme for energy states but quantum mechanics involves a discrete numerical scheme for energy states.

Geometry is another key component in modeling physical situations and can apply to either purely spatial parameters or to a combination of spatial, temporal or other parameters. For example, the forces acting on a charged object must be dealt with geometrically when a net force is to be determined. The force due to gravity, the force due to electrical attraction or repulsion and the force due to tension from a spring or string may all be involved.

These forces must be treated vectorially and an understanding of the geometrical relationships is fundamental to an understanding of a behavior of the system. If the system is not at steady state, time becomes a factor that can enter algebraically as it affects the geometrical properties of the system.

The particular approach to geometry, in terms of the system of coordinates, depends largely on the characteristics of the problem. Equations that are extremely complex in rectangular coordinates, for example, may be quite simple in spherical coordinates. Commensurately, problems that involve primarily planes, for example, are best treated in rectangular coordinates.

Thus, although algebra and geometry may seem to be simple aspects of mathematical modeling of physical situations, they are in fact highly nuanced and can add considerable complexity and detail to models.

Dimensional Analysis
Dimensional analysis is simply a technique in which the units of the variables in an equation are analyzed. It is often used by scientists and engineers to determine if a derived equation or computation is plausible. While it does not guarantee that a stated relationship is correct, it does tell us that the relationship is at least reasonable. When there is a mix of various physical quantities being equated in a relationship, the units of both sides of the equation must be the same. We will examine a simple example to demonstrate how dimensional analysis can be used:

The Ideal Gas Law is used to predict change in temperature, volume, or pressure of a gas. It is:
$$PV=nRT$$
Where P=pressure [Pa]
V=volume [m^3]
n=numbers of moles of gas [mol]
T=temperature [K]
R=the gas constant 8.314472 [$m^3 \cdot Pa \cdot K^{-1} \cdot mol^{-1}$]

Show that this formula is physically plausible using dimensional analysis.

Begin by substituting the units of each quantity into the equation:

$$[Pa] \times [m^3] = [mol] \left[\frac{[m^3] \times [Pa]}{[K] \times [mol]} \right] [K]$$

On the right-hand side of the equation, we can cancel [mol] and [K]:

$$[Pa] \times [m^3] = [m^3] \times [Pa]$$

Because this is a simple example, we can already see that the two sides of the equations have the same units. In more complicated equations, additional manipulation may be required to elucidate this fact. Now, we can confidently believe in the plausibility of this relationship.

TEACHER CERTIFICATION STUDY GUIDE

Skill 1.8 Interpreting results presented in different formats

Charts and Graphs

Scatterplots are typically shown on a Cartesian plane and are useful for demonstrating the relationship between two variables. A line chart (shown) is a special type of two-dimensional scatter plot in which the data points are connected with a line to make a trend more apparent.

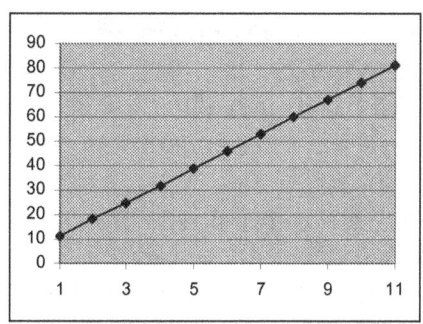

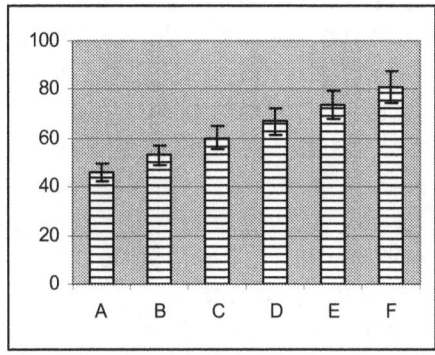

Bar charts can sometimes fill the same role as scatterplots but are better suited to show values across different categories or different experimental conditions (especially where those conditions are described qualitatively rather than quantified). Note the use of error bars in this example.

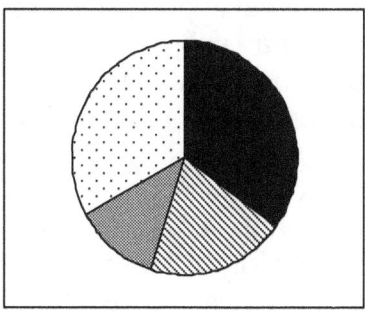

Finally, a pie chart is best used to present relative magnitudes or frequencies of several different conditions or events. They are most commonly used to show how various categories contribute to a whole.

Diagrams

Diagrams make it easy to visualize the connections and relationships between various elements. They may also be used to demonstrate temporal relationships. For example, diagrams can be used to illustrate the operation of an internal combustion engine or the complex biochemical pathways of an enzyme's action. The diagram shown is a simplified version of the carbon cycle.

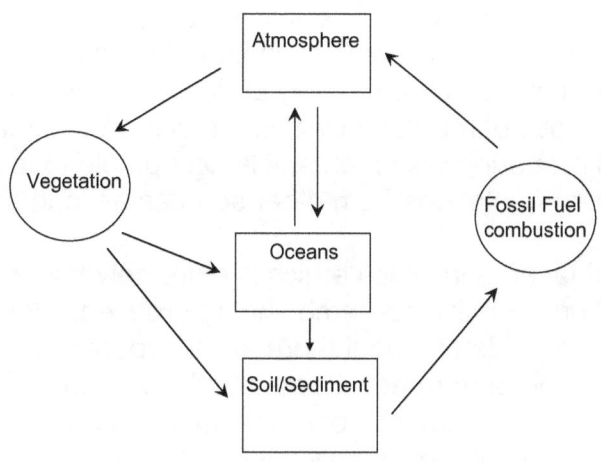

PHYSICS

Skill 1.9　Evaluating the validity of conclusions

Ethical behavior is critical at all points of scientific experimentation, interpretation, and communication. Both other scientists and the public in general need to be confident in the validity of scientific results and the fact that science is being performed in the service of public good.

The first aspect of this is that science is actually performed responsibly. It is always important that careful documents be kept of all experimental conditions and outcomes. In both commercial and academic research environments, scientists keep careful records in laboratory notebooks. These serve as primary documents bearing witness to new discoveries and developments. Additionally, it is of utmost importance that, when used, animal subjects are humanely treated and human subjects are fully aware of all relevant facts (informed consent). Most research and funding organizations have policies in place and review boards to monitor the use of animal and human subjects.

Secondly, the scientific findings must be accurately presented. This means that all relevant data is presented to clearly reflect experimental findings. Occasionally, a scientist will report only that data which agrees with the hypothesis he is putting forth. Alternatively, the data may be improperly statistically manipulated. These and other misrepresentations of experimental findings seriously impede the progress of science because they may provide misinformation about how the natural world operates.

Another aspect of the requirement to accurately present research is the need for the credit for work to be properly assigned. Whether in the development of technology or in pure research, there are both financial and scientific issues at stake. In academic and pure research settings findings are typically published in peer reviewed journals and student theses. Occasionally, cases of plagiarism or even the theft of experimental work occur. Plagiarism means that the work has been presented as original when, in fact, it incorporates the work of others. Both these occurrences are highly undesirable and may results in disciplinary action and/or loss of reputation within the scientific community. It should be noted that, when preparing documents, it is acceptable to include information and even direction quotations from other sources as long as they are always *properly cited*.

One final area in which ethics is extremely important is in deciding in what direction scientific research should proceed. Both scientists and the general public must determine if there are any possible investigations that *should* not actually be performed. Currently, this is especially important in medical research where appropriateness of research into human cloning, embryonic stems cells, and other topics are heavily debated.

Skill 1.10 Assessing the reliability of sources of information

There are many possible source of information about science. Their reliability and relevance are varied and each one may be particularly suited to a given circumstance.

Popular articles and books

When investigating an unfamiliar science-related topic, articles in popular science magazines, such as *Science*, *Popular Science* and even *National Geographic*, can be a helpful starting point. Such magazines often provide articles on scientific discoveries and ideas that cover the topic at an introductory level that is accessible to most readers. In addition to presenting some of the basic (and, sometimes, more advanced) concepts, these articles almost invariably also use at least some vocabulary associated with the topic. Many popular books also serve the same purpose as their counterparts in periodical literature, but provide a more thorough presentation of the topic. Although such popular articles and books may provide a good starting point when learning about some subject, they seldom offer a solid foundational or mathematical understanding. As a result, other sources must be consulted for more in-depth knowledge.

Textbooks

Student-oriented textbooks are a good resource for acquiring a more fundamental overview of a topic related to science. Textbooks regularly define critical terms and vocabulary, and very often include a bibliography that can be used as a starting point for more in-depth reading or research. Although textbooks often provide a reasonable theoretical foundation for understanding a particular subject, the discussion of relevant examples, current issues, and other specific information is often highly limited for the sake of brevity. Additionally, even classic textbooks that adequately cover certain topics may be out of date with respect to recent advances and controversies. As a result, textbooks are not an ideal source of information for highly specific or novel research. Some books that cover more specific topics are available and may be better resources than textbooks, however.

Peer-reviewed literature

The most up-to-date and informative literature dealing with specific issues and controversies in science is found in the peer-reviewed publications. These publications report new discoveries and findings and provide the most unadulterated view of the current status of a particular field of science. Due to the high-level technical nature of these publications, however, they are often inaccessible to readers not already familiar with at least the basics of the field. Also, knowledge of most of the technical vocabulary is assumed in journal articles. Nevertheless, with appropriate research, most papers can be unraveled and information can be gleaned from their pages. Furthermore, many of these technical articles have extensive bibliographies that can be used to do further background or related research.

Strategies for acquiring information

How one goes about studying or learning about a specific science-related topic or field depends largely on the specificity and technical level of the desired understanding, as well as on the background of the investigator. Someone new to a particular field may need to start at the level of a popular article or book and then build up, by way of textbooks and other middle-level resources, to peer-reviewed literature. Those with a solid background in a field who are seeking to do original research on a specific topic may find it sufficient to go directly to the peer-reviewed literature. The best strategy for learning about a science-related topic must take into account these and other considerations.

COMPETENCY 2.0 APPLY KNOWLEDGE OF METHODS AND EQUIPMENT USED IN SCIENTIFIC INVESTIGATIONS.

Skill 2.1 Includes selecting and using appropriate measurement devices and methods for collecting data

The approach to taking measurements for a particular experiment is determined by the particular phenomenon being measured, the context of measurement and the financial limitations imposed upon the experiment. Obviously, thermometers cannot be used for measuring distances, and the right equipment must be used for the appropriate measurement. Nevertheless, more subtle considerations abound, including the required accuracy and precision of measurements. If relatively coarse measurements are sufficient, there is often no need for sophisticated, expensive equipment.
For example, if a rough measurement of weight is required, a mechanical balance scale may be all that is required. Measurements of higher precision using an advanced digital scale may be gratuitous, especially in light of other inexact measurements that may be taken during the experiment.

The particular method used for measurement is determined in large part by the theory or hypothesis being tested. Measurements in the realm of particle physics, where hypotheses are largely based on a synthesis of quantum mechanics and special relativity, do not allow for the use of approaches to measurement that assume classical mechanics and electrodynamics. Instead, the particular method used must be based on more firmly established principles and concepts from the theory. Additionally, certain types of measurements must be approached indirectly. That is to say, empirical information about some parameters of a system may actually require measurement of different parameters from which measurements the desired parameter can be calculated. Such indirect measurements may be required for phenomena that are newly discovered, that are microscopic in scale or that require extremely high-precision results.

Skill 2.2 Evaluating the accuracy and precision of measurement and methods collecting data

Accuracy is a measure of how close to "correct" a measuring device or technique is. Precision is a measure of how similar repeated measurements from a given device or technique are. While the best devices or techniques will yield measurements that are both accurate and precise, it is possible to be accurate without being precise or to be precise without being accurate.

The classic analogy to demonstrate accuracy and precision is that of a bulls-eye. Accuracy alone is shown in the left example: the shots are all close to the center of the bulls-eye (the correct value). Precision alone is shown in the middle example: the shots are tightly clustered together. Both accuracy and precision are shown in the example on the right: the shots are tightly clustered near the center of the bulls-eye.

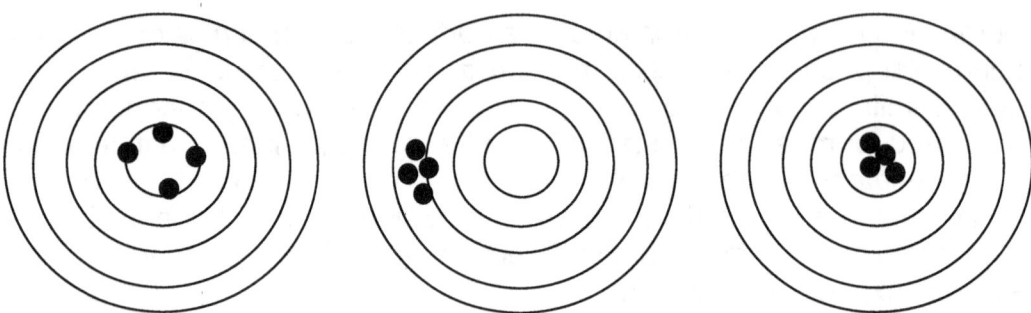

The accuracy of a technique or device can be determined in a straightforward manner. It is done by measuring a known quantity (a standard) and determining how close the value provided by the measuring device is. To determine precision, we must make multiple measurements of the same sample. The precision of an instrument is typically given in terms of its standard error or standard deviation. Precision is typically divided into reproducibility and repeatability. These concepts are subtly different and are defined as follows:

Repeatability: Variation observed in measurements made over a short period of time while trying to keep all conditions the same (including using the same instrument, the same environmental conditions, and the same operator)

Reproducibility: Variation observed in measurements taken over a long time period in a variety of different settings (different places and environments, using different instruments and operators)

Both repeatability and reproducibility can be estimated by taking multiple measurements under the conditions specified above. Using the obtained values, standard deviation can be calculated using the formula:

where σ = standard deviation
N = the number of measurements
x_i = the individual measured values
$\overline{x}$ = the average value of the measured quantity

$$\sigma = \sqrt{\frac{1}{N}\sum_{i=1}^{N}(x_i - \overline{x})^2}$$

To obtain a reliable estimate of standard deviation, N, the number of samples, should be fairly large. We can use statistical methods to determine a confidence interval on our measurements. A typical confidence level for scientific investigations is 90% or 95%.

Skill 2.3 Identifying procedures and sources of information related to the safe use, storage, and disposal of materials and equipment related to physics investigations measurement

Motion and forces: All stationary devices must be secured by C-clamps. Protective goggles must be used. Care must be taken at all times while knives, glass rods and heavy weights are used. Viewing a solar eclipse must always be indirect. When using model rockets, NASA's safety code must be implemented.

Heat: The master gas valve must be off at all times except while in use. Goggles and insulated gloves are to be used whenever needed. Never use closed containers for heating. Burners and gas connections must be checked periodically. Gas jets must be closed soon after the experiment is over. Fire retardant pads and quality glassware such as Pyrex must be used.

Pressure: While using a pressure cooker, never allow pressure to exceed 20 lb/square inch. The pressure cooker must be cooled before it is opened. Care must be taken when using mercury since it is poisonous. A drop of oil on mercury will prevent the mercury vapors from escaping.

Light: Broken mirrors or those with jagged edges must be discarded immediately. Sharp-edged mirrors must be taped. Spectroscopic light voltage connections must be checked periodically. Care must be taken while using ultraviolet light sources. Some students may have psychological or physiological reactions to the effects of strobe like (e.g. epilepsy).

Lasers: Direct exposure to lasers must not be permitted. The laser target must be made of non-reflecting material. The movement of students must be restricted during experiments with lasers. A number of precautions while using lasers must be taken – use of low power lasers, use of approved laser goggles, maintaining the room's brightness so that the pupils of the eyes remain small. Appropriate beam stops must be set up to terminate the laser beam when needed. Prisms should be set up before class to avoid unexpected reflection.

Sound: Fastening of the safety disc while using the high speed siren disc is very important. Teachers must be aware of the fact that sounds higher than 110 decibels will cause damage to hearing.

Radioactivity: The teacher must be knowledgeable and properly trained to handle the equipment and to demonstrate. Proper shielding of radioactive material and proper handling of material are absolutely critical. Disposal of any radioactive material must comply with the guidelines of NRC.

Chemicals: All laboratory solutions should be prepared as directed in the lab manual. Care should be taken to avoid contamination. All glassware should be rinsed thoroughly with distilled water before using and cleaned well after use. Unused solutions should be disposed of according to local disposal procedures. Chemicals should not be stored on bench tops or heat sources. They should be stored in groups based on their reactivity with one another and in protective storage cabinets. All containers within the lab must be labeled. Chemical waste should be disposed of in properly labeled containers. Waste should be separated based on their reactivity with other chemicals. Material Safety Data Sheets (MSDS) are available for every chemical substance directly from the company of acquisition or the internet.

It is important that teachers and educators follow these guidelines to protect the students and to avoid most of the hazards. They have a responsibility to protect themselves as well. There should be not any compromises in issues of safety.

Skill 2.4 Identifying hazards associated with laboratory practices and materials

Chemicals
The following chemicals are potential carcinogens and not allowed in school facilities: Acrylonitriel, Arsenic compounds, Asbestos, Bensidine, Benzene, Cadmium compounds, Chloroform, Chromium compounds, Ethylene oxide, Ortho-toluidine, Nickel powder, and Mercury.

Radiation
Proper shielding must be used while doing experiments with x-rays. All tubes that are used in a physics laboratory such as vacuum tubes, heat effect tubes, magnetic or deflection tubes must be checked and used for demonstrations by the teacher. Cathode rays must be enclosed in a frame and only the teacher should move them from their storage space. Students must watch the demonstration from at least eight feet away.

Electricity and Electrical Equipment
Active learning about electromagnetism may require the use of live electrical sources and many physics laboratories can be greatly enhanced with the use of electrical devices. However, care must always be taken when dealing with electricity. The danger in working with electrical device is two-fold as follows.

Risk of electrical shock due to:
    Exposed electrical contacts
    Poor or damaged insulation
    Poor or damaged equipment/cords
    Moisture in the area of electrical equipment

Risk of fire due to:
    Over-heating of equipment
    Overloaded circuits
    Faulty connections or short circuits
    Flammable substance in the area of electrical equipment

Adherence to the following rules will minimize the risk of electrical accidents:

1. Experiments should never be performed by a single person working alone.
2. Check all equipment, switches, and leads one by one. If missing insulation, frayed cords, blackening (due to arcing), bent or missing prongs are in evidence, do not use the equipment.
3. Always connect and turn the power supply on last. When concluding work, turn the power supply off first.
4. Do not run wires (or extension cords) over moving or rotating equipment, or on the floor, or string them across walkways from bench-to-bench.
5. Do not wear conductive watch bands or chains, finger rings, wrist watches, etc., and do not use metallic pencils, metal or metal edge rulers when working with exposed circuits.
6. If breaking an inductive circuit open, turn your face away to avoid danger from possible arcing.
7. If using electrolytic capacitors, always wait an appropriate period of time (usually approximately five time constants) for capacitor discharge before working on the circuit.
8. All conducting surfaces serving as "ground" should be connected together.
9. All instructional laboratories should be equipped with Ground Fault Current Interrupt (GFCI) circuit breakers. If breakers repeatedly trip and overload is not present, check for leakage paths to ground.
10. Use only dry hands and tools, work on a dry surface, and stand on a dry surface.
11. Never put conductive metal objects into energized equipment.
12. Always carry equipment by the handle and/or base, never by the cord.
13. Unplug cords by pulling on the plug, never on the cord.
14. Use extension cords only temporarily. The cord should be appropriately rated for the job.
15. Use extension cords with 3 prong plugs to ensure that equipment is grounded.
16. Never remove the grounding post from a 3 prong plug.
17. Do not overload extension cords, multi-outlet strips, or wall outlets.

Skill 2.5 Applying procedures for preventing accidents and dealing with emergencies

Safety is a learned behavior and must be incorporated into instructional plans. Measures of prevention and procedures for dealing with emergencies in hazardous situations have to be in place and readily available for reference. Copies of these must be given to all people concerned, such as administrators and students.

The single most important aspect of safety is planning and anticipating various possibilities and preparing for the eventuality. Any Physics teacher/educator planning on doing an experiment must try it before the students do it. In the event of an emergency, quick action can prevent many disasters. The teacher/educator must be willing to seek help at once without any hesitation because sometimes it may not be clear that the situation is hazardous and potentially dangerous.

There are a number of procedures to prevent and correct any hazardous situation. There are several safety aids available commercially such as posters, safety contracts, safety tests, safety citations, texts on safety in secondary classroom/laboratories, hand books on safety and a host of other equipment. Another important thing is to check the laboratory and classroom for safety and report it to the administrators before staring activities/experiments. It is important that teachers and educators follow these guidelines to protect the students and to avoid most of the hazards. They have a responsibility to protect themselves as well. There should be not any compromises in issues of safety.

All science labs should contain the following items of safety equipment.
-Fire blanket that is visible and accessible
-Ground Fault Circuit Interrupters (GCFI) within two feet of water supplies
-Signs designating room exits
-Emergency shower providing a continuous flow of water
-Emergency eye wash station that can be activated by the foot or forearm
-Eye protection for every student and a means of sanitizing equipment
-Emergency exhaust fans providing ventilation to the outside of the building
-Master cut-off switches for gas, electric and compressed air. Switches must have permanently attached handles. Cut-off switches must be clearly labeled.
-An ABC fire extinguisher
-Storage cabinets for flammable materials
-Chemical spill control kit
-Fume hood with a motor that is spark proof
-Protective laboratory aprons made of flame retardant material
-Signs that will alert potential hazardous conditions
-Labeled containers for broken glassware, flammables, corrosives, and waste.

Students should wear safety goggles when performing dissections, heating, or while using acids and bases. Hair should always be tied back and objects should never be placed in the mouth. Food should not be consumed while in the laboratory. Hands should always be washed before and after laboratory experiments. In case of an accident, eye washes and showers should be used for eye contamination or a chemical spill that covers the student's body. Small chemical spills should only be contained and cleaned by the teacher. Kitty litter or a chemical spill kit should be used to clean spill. For large spills, the school administration and the local fire department should be notified. Biological spills should only be handled by the teacher. Contamination with biological waste can be cleaned by using bleach when appropriate. Accidents and injuries should always be reported to the school administration and local health facilities. The severity of the accident or injury will determine the course of action to pursue.

COMPETENCY 3.0 UNDERSTAND THE DEVELOPMENT OF SCIENTIFIC THOUGHT AND INQUIRY

Skill 3.1 Includes demonstrating knowledge of the reliance of scientific investigations on empirical data, verifiable evidence, and logical reasoning

The scientific method is a logical set of steps that a scientist goes through to solve a problem. There are as many different scientific methods as there are scientists experimenting. However, there seems to be some pattern to their work. The scientific method is the process by which data is collected, interpreted and validated. While an inquiry may start at any point in this method and may not involve all of the steps here is the general pattern.

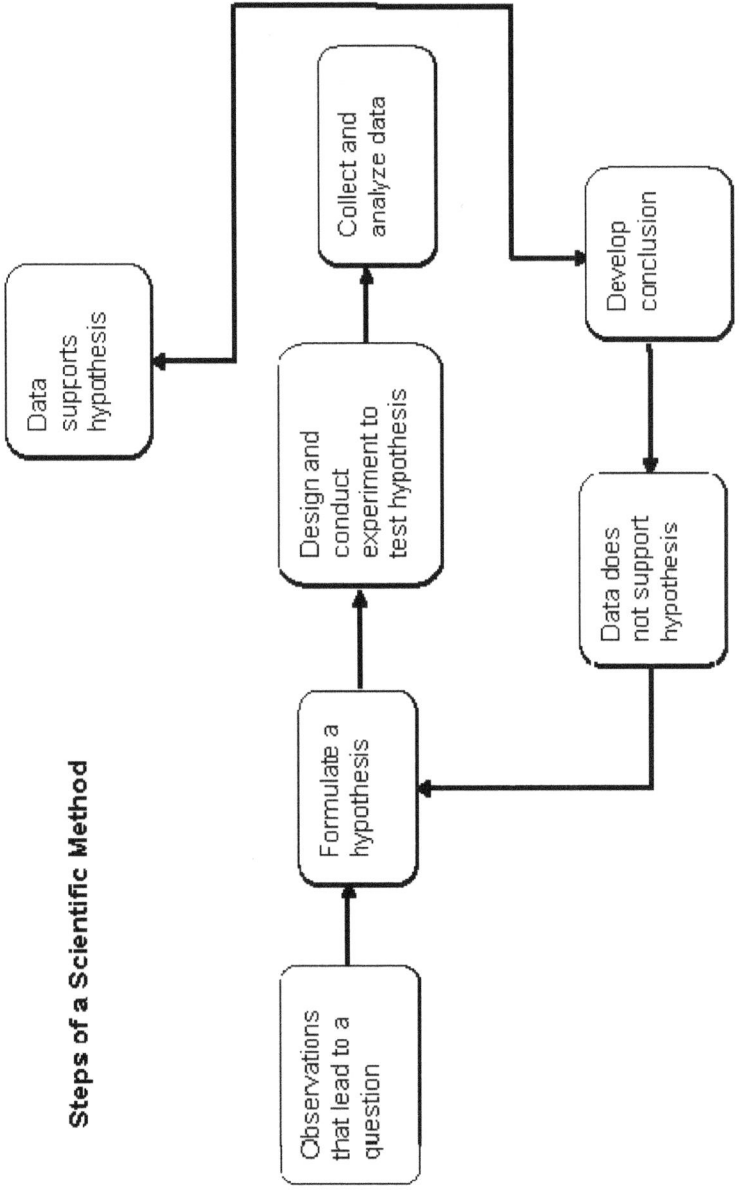

TEACHER CERTIFICATION STUDY GUIDE

Skill 3.2   **Recognizing the effect of researcher bias on scientific investigations and the interpretation of data**

Experimental bias occurs when a researcher favors one particular outcome over another in an experimental setup. In order to avoid bias, it is imperative to set up each experiment under exactly the same conditions, including a **control experiment**, an experiment with a known negative outcome. Additionally, in order to avoid experimental bias, a researcher must not "read" particular results into data.

An example of experimental bias can be seen in the example of the mouse in the maze experiment. In this example, a researcher is timing mice as they move through the maze towards a piece of cheese. The experiment relies on the mouse's ability to smell the cheese as it approaches. If one mouse chases a piece of cheddar cheese, while another chases Limburger, or so called "stinky" cheese, clearly the Limburger mouse has a huge advantage over the mouse chasing cheddar. To remove the experimental bias from this experiment, the same cheese should be used in both tests.

Skill 3.3   **Demonstrating an awareness of the contributions made to physics by individuals of diverse backgrounds and from different time periods**

Though people have tried to explain the phenomena of the natural world for millions of year, physics as we know it largely begins with the Greek scholars and continues with the Romans. Scientists in these civilizations were interested in the nature of matter, plant and animal life including the human body, simple mechanics, and the movement of heavenly bodies. Their scientific progress was somewhat impeded by the lack of accurate measuring devices and other advanced experimental equipment. Additionally, the loss of many written records (most notably as a result of the destruction of the Library of Alexandria) means that we do not have a full picture of their scientific knowledge.

It is known, however, that the Greeks did make many discoveries and formulate hypotheses that are still regarded as correct today. For instance, Greek mathematicians devised methods to calculate the area of three-dimensional objects that foreshadow integral calculus. Early Greek chemists speculated on the concept of a very tiny, indivisible form of matter (the atom) although they did not understand that each element was composed from a unique type of atom. Archimedes derived many correct principles of mechanics and hydrostatics (including his legendary "Eureka" moment that occurred when he noticed his own body displacing water in a bath). He also devised the Archimedes screw to transport water upward from rivers to farmland and formulated the law of the lever. The Greeks also possessed a surprising amount of knowledge about astronomy. Eratosthenes, for example, was able to determine that the Earth was a sphere and very accurately compute its circumference. Additionally,

Aristarchus of Samos deduced that the Earth rotated on an axis and orbited around the sun yearly. Scientific progress continued within the Roman Empire but their most important contributions to physics are related to their feats of engineering (roads, bridges, aqueducts, etc).

In the Western world, we tend to be most familiar with the early science and engineering of the Greeks and Romans. Unfortunately, the lack of written records often means that we know little about the science and technology of other ancient people, for instance, the native people of the Americas. However, there is a great deal of evidence that many other early civilizations were quite scientifically advanced. For instance, ancient Indian astronomers made many accurate calculations of the Earth's size and had a heliocentric model of the solar system. They also had a fairly advanced understanding of chemistry and chemical reactions, including a rudimentary notion of the atom. Indian mathematicians were particularly advanced and possessed knowledge of trigonometric functions and the concept of zero. Chinese scientists were similarly advanced in the fields of mathematics and astronomy.

The Chinese were particularly gifted inventors and engineers. They were early developers of many technologies including abacuses, sundials, gunpowder, compasses, wheelbarrows, suspension bridges, parachutes, propellers, crossbows, and processes for papermaking and printing. Unlike the Indian and Chinese scientists, Muslim scholars had a relatively high amount of interaction with their counterparts in the Western World. Geography and cultural forces led to much cross-over with their early Greek and Roman counterparts. Persian scientists invented both algebra and the astrolabe which was used by scientists in the East and West to advance the study of the heavenly bodies. The Iraqi physicist Alhazen studied and then improved upon Ptolemy's theory of light refraction. His early discoveries in the field of optics are still used today. Perhaps the greatest contribution of early Islamic scientists was their insistence on experimentation which was instrumental in developing the modern scientific method.

Skill 3.4　Recognizing the dynamic nature of scientific knowledge, including ways in which scientific knowledge changes

Scientific knowledge is based on a firm foundation of observation. Though mathematics and logic play a major role in defining and deducing scientific theories, ultimately even the most beautiful and intricate theory has to win the support of experiment. A single observation that contradicts an established theory can bring the whole edifice down if confirmed and reproduced. Thus scientific knowledge can never be totally certain and is always open to change based on some new evidence.

Sometimes advanced measuring devices and new equipment make it possible for scientists to detect phenomenon that no one had noted before. Nothing seemed more certain than classical Newtonian physics which explained everything from the motion of the planets to the behavior of earthly objects. At the end of the nineteenth century Lord Kelvin expressed the opinion that physics was complete except for the existence of "two small clouds"; the null result of the Michelson-Morley experiment and the failure of classical physics to predict the spectral distribution of blackbody radiation. The "two small clouds" turned out to be far more significant than Lord Kelvin could have imagined and led to the birth of relativity and quantum theory both of which totally changed the way we see the nature of reality.

If scientific knowledge is not inviolable, what keeps it from being vulnerable to challenge from anybody who thinks they have evidence to contradict a theory? Even though scientific knowledge is not sacred, the scientific process is. No observation is considered valid unless it can be reproduced by another scientist working independently under the same conditions. The peer-review process ensures that all results reported by a scientist undergo strict scrutiny by others working in the same field. Thus it is the integrity of the scientific process that keeps scientific knowledge, despite its openness to change, firmly grounded in objectivity and logic.

TEACHER CERTIFICATION STUDY GUIDE

COMPETENCY 4.0 **UNDERSTAND THE RELATIONSHIPS OF PHYSICS TO TECHNOLOGICAL AND SOCIETAL ISSUES, BOTH CONTEMPORARY AND HISTORICAL**

Skill 4.1 **Recognizing the relationships between science and technology**

Science and technology have the center-stage in our daily lives. More and more, it is becoming impossible for people in developed societies to exist without the necessities (e.g. cell phone, home appliances) and conveniences (e.g. satellite TV) afforded by technology. In fact, every day things that used to be conveniences are becoming necessities. Apart from the things that science and technology provide for us, they also represent a mind set and way of thinking such as the application of objectivity or rational thinking in evaluating events and options in our lives. Here are some of the ways in which science and technology are applied in our daily lives:

Health care: In this area, we can see many of the fruits of science and technology in nutrition, genetics, and the development of therapeutic agents. We can see an example of the adaptation of organisms in the development of resistant strains of microbes in response to use of antibiotics. Organic chemistry and biochemistry have been exploited to identify therapeutic targets and to screen and develop new medicines. Advances in molecular biology and our understanding of inheritance have led to the development of genetic screening and allowed us to sequence the human genome.

Environment: There are two broad happenings in environmental science and technology. First, there are many studies being conducted to determine the effects of changing environmental conditions and pollutions. New instruments and monitoring systems have increased the accuracy of these results. Second, advances are being made to mitigate the effects of pollution, develop sustainable methods of agriculture and energy production, and improve waste management.

Agriculture: Development of new technology in agriculture is particularly important as we strive to feed more people with less arable land. Again we see the importance of genetics in developing hybrids that have desirable characteristics. New strains of plants and farming techniques may allow the production of more nutrient rich food and/or allow crops to be grown successfully in harsh conditions. However, it is also important to consider the environmental impact of transgenic species and the use of pesticides and fertilizers. Scientific reasoning and experimentation can assist us in ascertaining the real effect of modern agricultural practices and ways to minimize their impact.

Information technology: The internet has become a new space in our lives. It is the global commons. It is where we conduct business, meet friends, obtain information and find entertainment. It affects all areas of human endeavor by allowing people worldwide to communicate easily and share ideas. It has also spawned a variety of new businesses.

With advances in technology come those in society who oppose it. Ethical questions come into play when discussing issues such as stem cell research or animal research for example. Does it need to be done? What are the effects on humans and animals? The answers to these questions are not always clear and often dependent on circumstance. Is the scientific process of organizing and weighing evidence applicable to these questions or do they lie beyond the domain of science and technology? These are the difficult issues that we have to face as technology moves forward.

Skill 4.2   Analyzing political and social factors that influence developments in physics, including current issues and controversies related to physics research and technology

Advances in science and technology create challenges and ethical dilemmas that national governments and society in general must attempt to solve. Local, state, national, and global governments and organizations must increasingly consider policy issues related to science and technology. For example, local and state governments must analyze the impact of proposed development and growth on the environment. Governments and communities must balance the demands of an expanding human population with the local ecology to ensure sustainable growth. Genetic research and manipulation, antibiotic resistance, stem cell research, and cloning are but a few of the issues facing national governments and global organizations today.

In all cases, policy makers must analyze all sides of an issue and attempt to find a solution that protects society while limiting scientific inquiry as little as possible. For example, policy makers must weigh the potential benefits of stem cell research, genetic engineering, and cloning (e.g. medical treatments) against the ethical and scientific concerns surrounding these practices. Many safety concerns have answered by strict government regulations. The FDA, USDA, EPA, and National Institutes of Health are just a few of the government agencies that regulate pharmaceutical, food, and environmental technology advancements

Scientific and technological breakthroughs greatly influence other fields of study and the job market as well. Advances in information technology have made it possible for all academic disciplines to utilize computers and the internet to simplify research and information sharing. In addition, science and technology influence the types of available jobs and the desired work skills.

For example, machines and computers continue to replace unskilled laborers and computer and technological literacy is now a requirement for many jobs and careers. Finally, science and technology continue to change the very nature of careers. Because of science and technology's great influence on all areas of the economy, and the continuing scientific and technological breakthroughs, careers are far less stable than in past eras. Workers can thus expect to change jobs and companies much more often than in the past.

Because people often attempt to use scientific evidence in support of political or personal agendas, the ability to evaluate the credibility of scientific claims is a necessary skill in today's society. The media and those with an agenda to advance often overemphasize the certainty and importance of experimental results. One should question any scientific claim that sounds fantastical or overly certain. Scientific, peer-reviewed journals are the most accepted source for information on scientific experiments and studies. Knowledge of experimental design and the scientific method is important in evaluating the credibility of studies. For example, one should look for the inclusion of control groups and the presence of data to support the given conclusions.

Skill 4.3 Evaluating the credibility of scientific claims made in various forums

Because people often attempt to use scientific evidence in support of political or personal agendas, the ability to evaluate the credibility of scientific claims is a necessary skill in today's society. In evaluating scientific claims made in the media, public debates, and advertising, one should follow several guidelines.

First, scientific, peer-reviewed journals are the most accepted source for information on scientific experiments and studies. One should carefully scrutinize any claim that does not reference peer-reviewed literature.

Second, the media and those with an agenda to advance (advertisers, debaters, etc.) often overemphasize the certainty and importance of experimental results. One should question any scientific claim that sounds fantastical or overly certain.

Finally, knowledge of experimental design and the scientific method is important in evaluating the credibility of studies. For example, one should look for the inclusion of control groups and the presence of data to support the given conclusions.

Also see **Skill 1.10** for more information on the various sources of scientific information.

COMPETENCY 5.0 UNDERSTAND INTERRELATIONSHIPS AMONG THE PHYSICAL, LIFE, AND EARTH/SPACE SCIENCES

Skill 5.1 Includes recognizing major unifying themes and concepts that are common to the various scientific disciplines

Math, science, and technology all have common themes in how they are applied and understood. Here are some of the fundamental concepts:

Systems
Because the natural world is so complex, the study of science involves the organization of items into smaller groups based on interaction or interdependence. These groups are known as systems. Systems consist of many separate parts interacting in specific ways to form a whole. It is these interactions that truly define the system. The complete system then has its own characteristics that go beyond the simple collection of its components (i.e., "the whole is more than the sum of the parts"). Natural phenomena and complex technologies can almost always be represented as systems. Examples of systems are the solar system, cardiovascular system, Newton's laws of force and motion, and the laws of conservation.

Models
Science and technology employ models to help simplify concepts. Models can be actual, small, physical mock-ups, mathematical equations, or diagrams that represent the fundamental relationships being studied. Models allow us to gain an understanding of these relationships and to make predictions. Similarly, diagrams, graphs, and charts are often employed to make these phenomena more readily understandable in a visual way.

Change and equilibrium
Another common theme among these three areas is the alternation between change and stability. These alternations occur in natural systems, which typically follow a pattern in which variation is introduced and then equilibrium is restored. Equilibrium is a state in which forces are balanced, resulting in stability. Static equilibrium is stability due to a lack of changes and dynamic equilibrium is stability due to a balance between opposite forces. Similarly, many technologies involve either creation or control of change. The process of change over a long period of time is known as evolution. While biological evolution is the most common example, one can also classify technological advancement, changes in the universe, and changes in the environment as evolution.

TEACHER CERTIFICATION STUDY GUIDE

Scale
In science and technology, we must deal with quantities that have vastly different magnitudes. It is important to understand the relationships between such very different numbers. Specifically, it must be recognized that behavior, and even the laws of physics, may change with scale. Some relationships, for instance, the effect of friction on speed, are only valid over certain size scales. When developing new technology, such as nano-machines, we must keep in mind the importance of scale.

Form and function
Form and function are properties of systems that are closely related. The function of an object usually dictates its form and the form of an object usually facilitates its function. For example, the form of the heart (e.g. muscle, valves) allows it to perform its function of circulating blood through the body. The idea of function dictating the form is also used in architecture.

Skill 5.2 Describing the integration and interdependence of the sciences

The laws of physics are those that govern all parts of the natural world. The laws of physics reveal the root explanation for every geological, chemical, biological, and astronomical phenomenon. Let's examine how knowledge of the different aspects of physics is useful in understanding phenomenon from each field.

Geology
The study of the physical laws that affect the earth and its processes is known as geophysics. Thermodynamics is needed to understand how heat moves between the core, mantle, and crust of the earth. Additionally, thermodynamics, along with the study of electromagnetism, is needed to comprehend and predict meteorological events. Mechanics is required to analyze the movement of the tectonic plates, earthquakes, and other seismic activity. Mechanics, particularly those of waves, is also key in the study of hydrology.

Chemistry
Physical chemistry is the branch of chemistry concerned with physical behavior of molecules and the physical aspects of chemical reactions. Concepts from modern physics including nuclear physics and quantum and statistical mechanics are of key importance in this field. Thermodynamics and basic mechanics are also useful in understanding the progress of chemical reactions. Chemists have also made extensive use of the laws of electromagnetism in designing analysis tools such as NMR spectrometers.

Biology

The application of physical theories and methods in biological systems is known as biophysics. This field is largely devoted to the micromechanics of the cell, particularly the cytoskeleton and the physical structure of proteins. However, understanding larger biological systems also requires an understanding of physics. For instance, thermodynamics is needed to analyze heat transfer in warm and cold blooded animals. Mechanics is necessary to understand and develop therapies for the musculoskeletal system.

Astronomy

Astrophysics is the study of the physics of the universe. Mechanics is important in predicting the motion of planets, comets, and other celestial bodies. Thermodynamics is used in studying the transfer of energy and the heating and cooling of planets and stars. Optics and an understanding of sound and light are also necessary, particularly for the proper use of telescopes and other instruments used for studying the universe.

Skill 5.3 Interdisciplinary connections among the sciences and their applications in real-world contexts

Science has increasingly become a collaborative and interdisciplinary effort. Engineers, clinicians, and scientists may not be familiar with advances in far reaching fields unless new developments are announced to a broad community of scientists. When this happens, it opens the doors for an exchange of information that can be enormously helpful in both investigating and solving problems in science and technology. Consider, for instance, an exchange of information between physicists and biologists. The physicists may suggest the use of high powered microscopes and lasers for study of sub-cellular structures. Likewise, the biologist's knowledge of transport proteins could inform the physicist's design of new nano-machines.

It is also important that the public in general be kept informed of scientific progress. This is of key importance in any democratic society because all people together must decide the direction of the government. New scientific discoveries can have huge implications for medical, social, and environmental issues which should be of interest to all voters. Additionally, when people are made familiar with the advances made in science, they may be more willing to see the need for public funding which supports continued scientific progress.

TEACHER CERTIFICATION STUDY GUIDE

SUBAREA II.   **MECHANICS**

COMPETENCY 6.0   ANALYZE MOTION IN ONE AND TWO DIMENSIONS

Skill 6.1   Analyzing information related to displacement, speed, velocity, and acceleration presented in one or more representations

Kinematics is the part of mechanics that seeks to understand the motion of objects, particularly the relationship between position, velocity, acceleration and time.

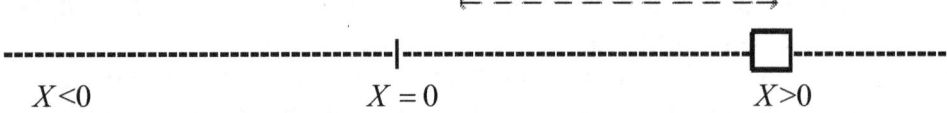

$$X<0 \qquad X=0 \qquad X>0$$

The above figure represents an object and its displacement along one linear dimension.

First we will define the relevant terms:

1. Position or Distance is usually represented by the variable *x*. It is measured relative to some fixed point or datum called the origin in linear units, meters, for example.

2. Displacement is defined as the change in position or distance which an object has moved and is represented by the variables D, d or Δx. Displacement is a vector with a magnitude and a direction.

3. Velocity is a vector quantity usually denoted with a V or v and defined as the rate of change of position. Typically units are distance/time, m/s for example. Since velocity is a vector, if an object changes the direction in which it is moving it changes its velocity even if the speed (the scalar quantity that is the magnitude of the velocity vector) remains unchanged.

     i) Average velocity: $\vec{v} \equiv \dfrac{\Delta d}{\Delta t} = d_1 - d_0 / t_1 - t_0$

         The ratio $\Delta d / \Delta t$ is called the average velocity. Average here denotes that this quantity is defined over a period Δt.

     ii) Instantaneous velocity is the velocity of an object at a particular moment in time. Conceptually, this can be imagined as the extreme case when Δt is infinitely small.

5. Acceleration represented by *a* is defined as the rate of change of velocity and the units are m/s^2. Both an average and an instantaneous acceleration can be defined similarly to velocity.

From these definitions we develop the kinematic equations. In the following, subscript i denotes initial and subscript f denotes final values for a time period. Acceleration is assumed to be constant with time.

$$v_f = v_i + at \qquad (1)$$

$$d = v_i t + \frac{1}{2}at^2 \qquad (2)$$

$$v_f^2 = v_i^2 + 2ad \qquad (3)$$

$$d = \left(\frac{v_i + v_f}{2}\right)t \qquad (4)$$

Example:
Leaving a traffic light a man accelerates at 10 m/s². a) How fast is he going when he has gone 100 m? b) How fast is he going in 4 seconds? C) How far does he travel in 20 seconds.

Solution:
a) Use equation 3. He starts from a stop so v_i=0 and v_f^2=2 x 10m/s² x 100m=2000 m²/s² and v_f=45 m/s.
b) Use equation 1. Initial velocity is again zero so v_f=10m/s² x 4s=40 m/s.
c) Use equation 2. Since initial velocity is again zero, d=1/2 x 10 m/s² x (20s)²=2000 m

The relationship between time, position or distance, velocity and acceleration can be understood conceptually by looking at a graphical representation of each as a function of time. Simply, the velocity is the slope of the position vs. time graph and the acceleration is the slope of the velocity vs. time graph. If you are familiar with calculus then you know that this relationship can be generalized: velocity is the first derivative and acceleration the second derivative of position.

Here are three examples:

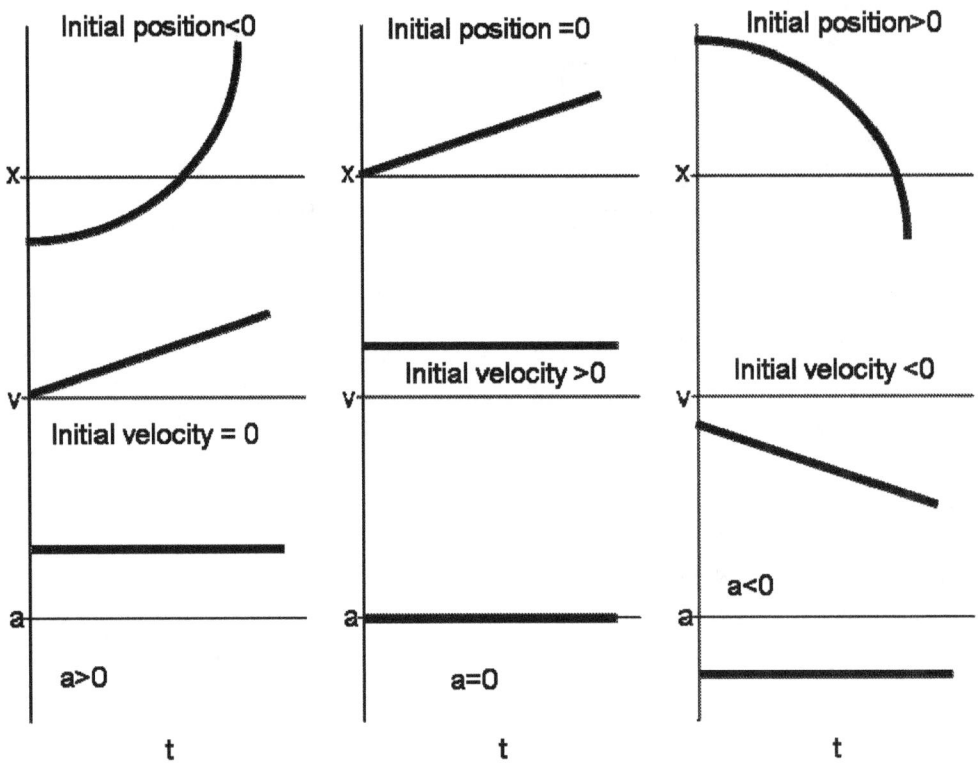

There are three things to notice:

1) In each case acceleration is constant. This isn't always the case, but a simplification for this illustration.

2) A non-zero acceleration produces a position curve that is a parabola.

3) In each case the initial velocity and position are specified separately. The acceleration curve gives the shape of the velocity curve, but not the initial value and the velocity curve gives the shape of the position curve but not the initial position.

Skill 6.2 Solving problems involving constant acceleration

Simple problems involving distance, displacement, speed, velocity, and constant acceleration can be solved by applying kinematics equations. The following steps should be employed to simplify a problem and apply the proper equations:

1. Create a simple diagram of the physical situation.
2. Ascribe a variable to each piece of information given.
3. List the unknown information in variable form.
4. Write down the relationships between variables in equation form.
5. Substitute known values into the equations and use algebra to solve for the unknowns.
6. Check your answer to ensure that it is reasonable.

Example:
A man in a truck is stopped at a traffic light. When the light turns green, he accelerates at a constant rate of 10 m/s².
a) How fast is he going when he has gone 100 m?
b) How fast is he going after 4 seconds?
c) How far does he travel in 20 seconds?

Solution:
We first construct a diagram of the situation. In this example, the diagram is very simple, only showing the truck accelerating at the given rate. Next we define variables for the known quantities (these are noted in the diagram): a=10 m/s²; v$_i$=0 m/s

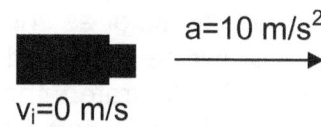

Now we will analyze each part of the problem, continuing with the process outlined above.

For part **a)**, we have one additional known variable: d=100 m
The unknowns are: v$_f$ (the velocity after the truck has traveled 100m)
Equation (3) will allow us to solve for v$_f$, using the known variables:

$$v_f^2 = v_i^2 + 2ad$$

$$v_f^2 = (0m/s)^2 + 2(10m/s^2)(100m) = 2000\frac{m^2}{s^2}$$

$$v_f = 45\frac{m}{s}$$

We use this same process to solve part **b)**. We have one additional known variable: t=4 seconds. The unknowns are: v$_f$ (the velocity after the truck has traveled for 4 seconds) Thus, we can use equation (1) to solve for v$_f$:

$$v_f = v_i + at$$

$$v_f = 0m/s + (10m/s^2)(4s) = 40m/s$$

For part **c)**, we have one additional known variable: t= 20 s
The unknowns are: d (the distance after the truck has traveled for 20 seconds)
Equation (2) will allow us to solve this problem:

$$d = v_i t + \frac{1}{2}at^2$$

$$d = (0m/s)(20s) + \frac{1}{2}(10m/s^2)(20s)^2 = 2000m$$

Finally, we consider whether these solutions seem physically reasonable. In this simple problem, we can easily say that they do.

Skill 6.3   Applying principles of trigonometry and properties of vectors to analyze motion in two dimensions

Vector space is a collection of objects that have **magnitude** and **direction**. They may have mathematical operations, such as addition, subtraction, and scaling, applied to them. Vectors are usually displayed in boldface or with an arrow above the letter. They are usually shown in graphs or other diagrams as arrows. The length of the arrow represents the magnitude of the vector while the direction in which the arrow points shows the vector direction.

To **add two vectors** graphically, the base of the second vector is drawn from the point of the first vector as shown below with vectors **A** and **B**. The sum of the vectors is drawn as a dashed line, from the base of the first vector to the tip of the second. As illustrated, the order in which the vectors are connected is not significant as the endpoint is the same graphically whether **A** connects to **B** or **B** connects to **A**. This principle is sometimes called the parallelogram rule.

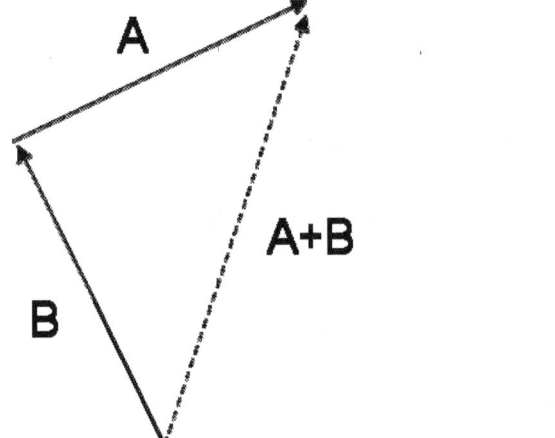
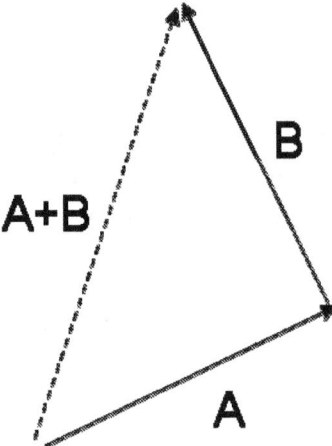

If more than two vectors are to be combined, additional vectors are simply drawn in accordingly with the sum vector connecting the base of the first to the tip of the final vector.

Subtraction of two vectors can be geometrically defined as follows. To subtract **A** from **B**, place the ends of **A** and **B** at the same point and then draw an arrow from the tip of **A** to the tip of **B**. That arrow represents the vector **B-A**, as illustrated below:

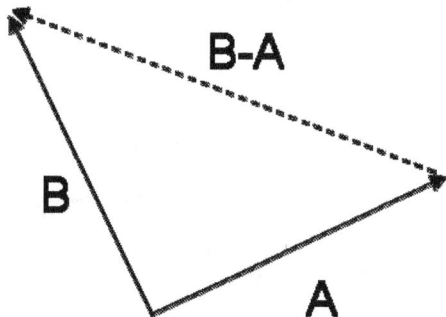

To add two vectors without drawing them, the vectors must be broken down into their orthogonal components using sine, cosine, and tangent functions. Add both x components to get the total x component of the sum vector, then add both y components to get the y component of the sum vector. Use the Pythagorean Theorem and the three trigonometric functions to the get the size and direction of the final vector.

Example: Here is a diagram showing the x and y-components of a vector D1:

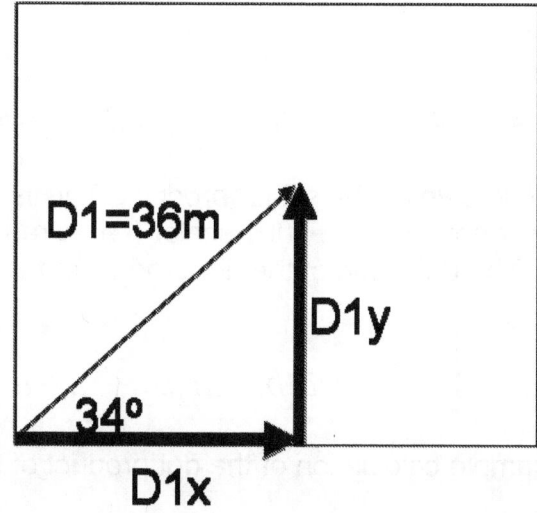

Notice that the x-component D1x is adjacent to the angle of 34 degrees.

Thus D1x=36m (cos34) =29.8m

The y-component is opposite to the angle of 34 degrees.

Thus D1y =36m (sin34) = 20.1m

A second vector D2 is broken up into its components in the diagram below using the same techniques. We find that D2y=9.0m and D2x=-18.5m.

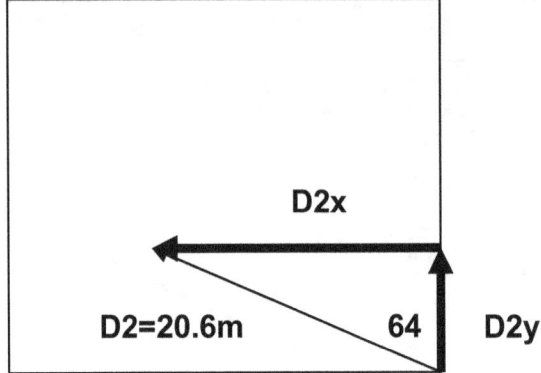

Next we add the x components and the y components to get

DTotal x =11.3 m   *and*   DTotal y =29.1 m

Now we have to use the Pythagorean theorem to get the total magnitude of the final vector. And the arctangent function to find the direction. As shown in the diagram below.

DTotal=31.2m

tan θ= DTotal y / DTotal x = 29.1m / 11.3 =2.6   θ=69 degrees

The dot product is also known as the scalar product. This is because the dot product of two vectors is not a vector, but a scalar (i.e., a real number without an associated direction). The definition of the dot product of the two vectors **a** and **b** is:

$$a \bullet b = \sum_{i=1}^{n} a_i b_i = a_1 b_1 + a_2 b_2 + ... + a_n b_n$$

The following is an example calculation of the dot product of two vectors:

$$[1\ 3\ -5] \cdot [4\ -2\ -2] = (1)(4) + (3)(-2) + (-5)(-2) = 8$$

Note that the product is a simple scalar quantity, not a vector. The dot product is commutative and distributive.

Unlike the dot product, the cross product does return another vector. The vector returned by the cross product is orthogonal to the two original vectors. The cross product is defined as:

$$\mathbf{a} \times \mathbf{b} = \mathbf{n} |\mathbf{a}| |\mathbf{b}| \sin \theta$$

where n is a unit vector perpendicular to both **a** and **b** and θ is the angle between **a** and **b**. In practice, the cross product can be calculated as explained below:

Given the orthogonal unit vectors **i**, **j**, and **k**, the vector **a** and **b** can be expressed:

$$\mathbf{a} = a_1\mathbf{i} + a_2\mathbf{j} + a_3\mathbf{k}$$
$$\mathbf{b} = b_1\mathbf{i} + b_2\mathbf{j} + b_3\mathbf{k}$$

Then we can calculate that: $\mathbf{a} \times \mathbf{b} = \mathbf{i}(a_2 b_3) + \mathbf{j}(a_3 b_1) + \mathbf{k}(a_1 b_2) - \mathbf{i}(a_3 b_2) - \mathbf{j}(a_1 b_3) - \mathbf{k}(a_2 b_1)$

The cross product is anticommutative (that is, $\mathbf{a} \times \mathbf{b} = -\mathbf{b} \times \mathbf{a}$) and distributive over addition.

Two-Dimensional Motion
The most common example of an object moving in two dimensions is a projectile. A projectile is an object upon which the only force acting is gravity. Some examples:
  i) An object dropped from rest
  ii) An object thrown vertically upwards at an angle
  iii) A canon ball

Once a projectile has been put in motion (say, by a canon or hand) the only force acting it is gravity, which near the surface of the earth implies it experiences a=g=9.8m/s².

This is most easily considered with an example such as the case of a bullet shot horizontally from a standard height at the same moment that a bullet is dropped from exactly the same height. Which will hit the ground first? If we assume wind resistance is negligible, then the acceleration due to gravity is our only acceleration on either bullet and we must conclude that they will hit the ground at the same time. The horizontal motion of the bullet is not affected by the downward acceleration.

Example:
I shoot a projectile at 1000 m/s from a perfectly horizontal barrel exactly 1 m above the ground. How far does it travel before hitting the ground?

Solution:
First figure out how long it takes to hit the ground by analyzing the motion in the vertical direction. In the vertical direction, the initial velocity is zero so we can rearrange kinematic equation 2 from the previous section to give:

$t = \sqrt{\dfrac{2d}{a}}$. Since our displacement is 1 m and a=g=9.8m/s^2, t=0.45 s.

Now use the time to hitting the ground from the previous calculation to calculate how far it will travel horizontally. Here the velocity is 1000m/s and there is no acceleration. So we simple multiply velocity with time to get the distance of 450m.

Circular Motion
Motion on an arc can also be considered from the view point of the kinematic equations. As pointed out earlier, displacement, velocity and acceleration are all vector quantities, i.e. they have magnitude (the speed is the magnitude of the velocity vector) and direction. This means that if one drives in a circle at constant speed one still experiences an acceleration that changes the direction. We can define a couple of parameters for objects moving on circular paths and see how they relate to the kinematic equations.

Tangential speed: The tangent to a circle or arc is a line that intersects the arc at exactly one point. If you were driving in a circle and instantaneously moved the steering wheel back to straight, the line you would follow would be the tangent to the circle at the point where you moved the wheel. The tangential speed then is the instantaneous magnitude of the velocity vector as one moves around the circle.

Tangential acceleration: The tangential acceleration is the component of acceleration that would change the tangential speed and this can be treated as a linear acceleration if one imagines that the circular path is unrolled and made linear.

Centripetal acceleration: Centripetal acceleration corresponds to the constant change in the direction of the velocity vector necessary to maintain a circular path. Always acting toward the center of the circle, centripetal acceleration has a magnitude proportional to the tangential speed squared divided by the radius of the path.

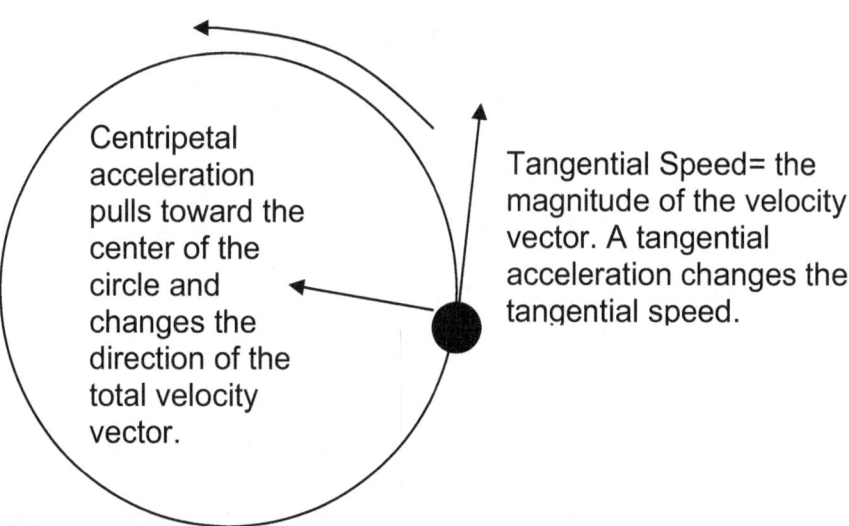

Skill 6.4 Applying calculus to analyze motion in one dimension

The widespread use of calculus, as applied to problems of physics, can be largely credited to Sir Isaac Newton. Gottfried Wilhelm Leibniz is credited with having simultaneously developed calculus although his work did not have as much physics-oriented character as did Newton's work. The integral and differential, though abstract in and of themselves, can be related to various physical situations and can provide extremely useful analytical tools.

In simple terms, the derivative of a function is essentially a slope or a rate of change of that function at each point in a defined range. Thus, the derivative of a function that relates the position of a particle to time is the rate of change of the position, or speed. The derivative of a function over some range is itself a function, and can also be differentiated. Following the previous example, the derivative of the speed of a particle is the rate of change of the speed, or the acceleration. Thus, derivatives can be used to determine additional information about a system based on other information or observations.

In cases where a function involves more than one variable, partial derivatives may be involved. These are simply treated as derivatives in terms of the specified variable while assuming that all other variables remain constant.

Partial derivatives are a crucial component of operators such as the gradient, which, in part, determines the direction (in multiple dimensions) of the greatest increase of a function.

The integral, also aptly named the antiderivative, is the area under the curve of a function in a given range. The function for the speed of a car can be integrated over time to get distance. This is a generalization of the case of constant speed where the speed is simply multiplied by the time traveled to get the distance traveled. The integral performs this multiplication operation over infinitesimally small increments of time thus allowing for calculation in the case of time-varying speed. Integrals may be performed for a single variable even in the case of a function containing several variables. The approach to integration in this case is the same as the approach to partial derivatives: all variables not being integrated are treated as constant.

Vector calculus is often necessary for physical situations where the directions of certain parameters, in addition to the magnitudes, are involved. Disciplines such as classical electrodynamics, for example, must largely be treated with vector calculus. For vector calculus, the choice of coordinate system becomes slightly more complicated, as in some cases the unit vectors are variable. The rectangular coordinate system involves unit vectors that are constant, making it a preferred system of coordinates so long as the expressions are not overly difficult. In these cases it may be beneficial to use other coordinate systems such as cylindrical or spherical.

Trends and patterns in a set of data are most easily identified when the data is displayed graphically. Exact data values in tabular form are also used to calculate various features of the data set. Following are some aspects of data that are commonly analyzed in all scientific disciplines.

The **slope** or the **gradient** of a line is used to describe the rate of change of a variable with respect to another or, in calculus terms, the **derivative** of one variable with respect to another. In the set of examples shown below, the relationship between time, position or distance, velocity and acceleration can be understood conceptually by looking at a graphical representation of each as a function of time. The velocity is the slope of the position vs. time graph and the acceleration is the slope of the velocity vs. time graph.

Here are some things we notice by inspecting these graphs:

    1) In each case acceleration is constant.

    2) A non-zero acceleration produces a position curve that is a parabola.

    3) In each case the initial velocity and position are specified separately.

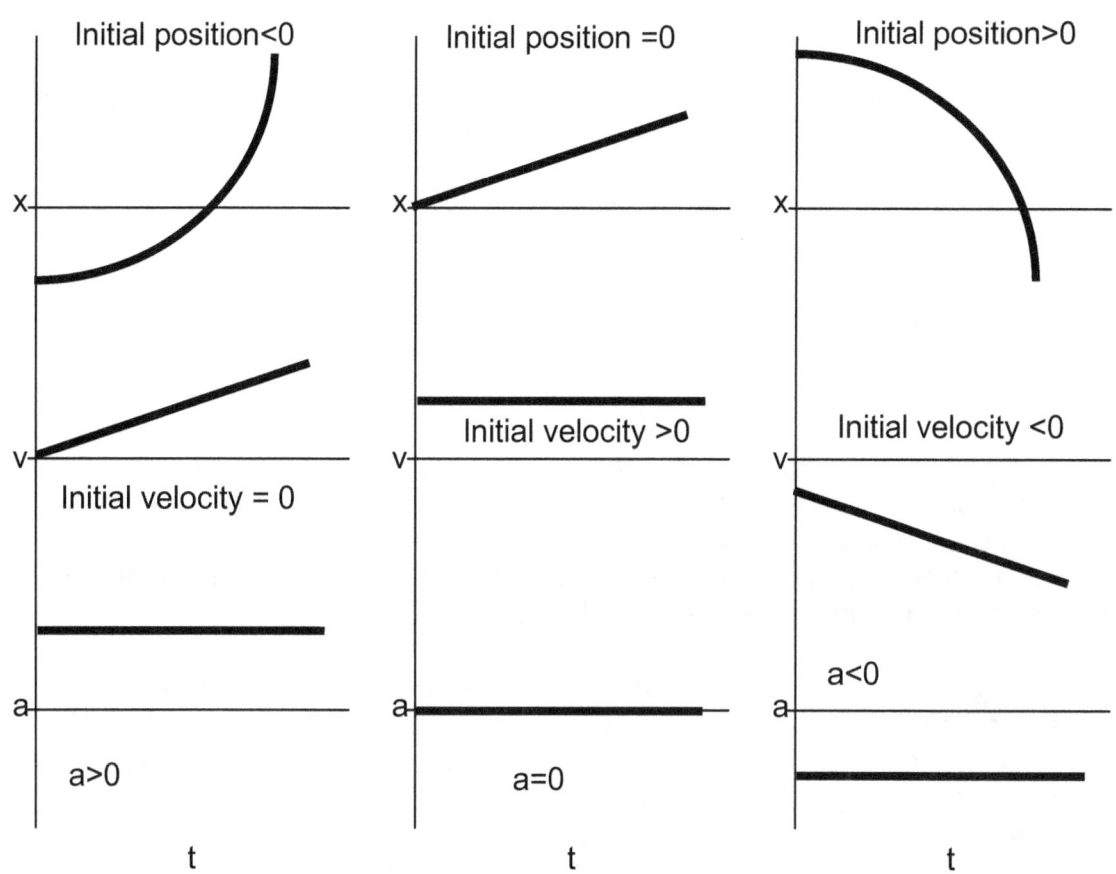

The acceleration curve gives the shape of the velocity curve, but not the initial value and the velocity curve gives the shape of the position curve but not the initial position.

The slope of a line may be calculated using the data values displayed in the graph and is given by
$$m = \frac{\Delta y}{\Delta x}$$

Given two points (x1, y1) and (x2, y2), the change in x from one to the other is x2 - x1, while the change in y is y2 - y1. Substituting both quantities into the above equation obtains the following:

$$m = \frac{y_2 - y_1}{x_2 - x_1}$$

For example: if a line runs through two points: P(1,2) and Q(13,8). By dividing the difference in y-coordinates by the difference in x-coordinates, one can obtain the slope of the line:

$$m = \frac{\Delta y}{\Delta x} = \frac{y_2 - y_1}{x_2 - x_1} = \frac{8 - 2}{13 - 1} = \frac{6}{12} = \frac{1}{2}$$

The slope is 1/2 = 0.5.

The slope of curved lines can be approximated by selecting x and y values that are very close together. In a curved region the slope changes along the curve.

If we let Δx and Δy be the x and y distances between two points on a curve, then Δy /Δx is the slope of a secant line to the curve.

For example, the slope of the secant intersecting
$y = x^2$ at (0,0) and (3,9)
is m = (9 - 0) / (3 - 0) = 3

By moving the two points closer together so that Δy and Δx decrease, the secant line more closely approximates a tangent line to the curve, and as such the slope of the secant approaches that of the tangent. In differential calculus, the derivative is essentially taking the change in y with respect to the change in x as the change in x approaches the limit of zero.

The **integral** is usually used to find a measure of totality such as area, volume, mass, or displacement when the rate of change is specified, as in any simple x-y graphical representation. The integral of a function is an extension of the concept of summing and is given by the area under a graphical representation of the function.

The simplest graph to analyze the area under is a flat horizontal line. As an example, let's say *f* is the constant function $f(x) = 3$ and we want to find the area under the graph from x= 0 to x=10. This is simply a rectangle 3 units high by 10 units long, or 30 units square. The same result can be found by integrating the function, though this is usually done for more complicated or smooth curves.

Let us imagine the curve of a function f(X) between X=0 and X=10. One way to approximate the area under the curve is to draw numerous rectangles under the curve of a given width, estimate their height and sum the area of each rectangle. We can say that the width of each rectangle is δX. But since the top of each column is not exactly straight, this is only an approximation. When we use integral calculus to determine an integral, we are taking the limit of δX approaching zero, so there will be more and more columns which are thinner and thinner to fill the space between X=0 and X=10. The top of each column then gets closer and closer to being a straight line and our expression for the area therefore gets closer and closer to being exactly right.

TEACHER CERTIFICATION STUDY GUIDE

COMPETENCY 7.0 UNDERSTAND NEWTON'S LAWS OF MOTION AND THE LAW OF UNIVERSAL GRAVITATION

Skill 7.1 Applying Newton's laws of motion, both descriptively and mathematically, in a variety of situations

Newton's first law of motion: "An object at rest tends to stay at rest and an object in motion tends to stay in motion with the same speed and in the same direction unless acted upon by an unbalanced force". Prior to Newton's formulation of this law, being at rest was considered the natural state of all objects, because at the earth's surface we have the force of gravity working at all times which causes nearly any object put into motion to eventually come to rest. Newton's brilliant leap was to recognize that an unbalanced force changes the motion of a body, whether that body begins at rest or at some non-zero speed.

We experience the consequences of this law everyday. For instance, the first law is why seat belts are necessary to prevent injuries. When a car stops suddenly, say by hitting a road barrier, the driver continues on forward until acted upon by a force. The seat belt provides that force and distributes the load across the whole body rather than allowing the driver to fly forward and experience the force against the steering wheel.

Example: A skateboarder is riding her skateboard down a road. The skateboard has a constant speed of 5 m/s. Then the skateboard hits a rock and stops suddenly. Since the rider has nothing to stop her when the skateboard stops, she will continue to travel at 5 m/s until she hits the ground.

Newton's second law of motion: "The acceleration of an object as produced by a net force is directly proportional to the magnitude of the net force, in the same direction as the net force, and inversely proportional to the mass of the object". In the equation form, it is stated as $F = ma$ force equals mass times acceleration. It is important to remember that this is the net force and that forces are vector quantities. Thus if an object is acted upon by 12 forces that sum to zero, there is no acceleration. Also, this law embodies the idea of inertia as a consequence of mass. For a given force, the resulting acceleration is proportionally smaller for a more massive object because the larger object has more inertia.

Example:
A ball is dropped from a building. The mass of the ball is 2 kg. The acceleration of the object is 9.8 m/s² (gravitational acceleration).

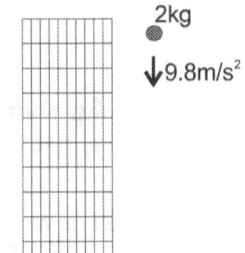

Therefore, the force acting on the ball is
$F = ma \Rightarrow F = 2 \text{ kg} \times 9.8 \text{ m/s}^2 \Rightarrow F = 19.6 \text{ N}$

Newton's third law of motion: "For every action, there is an equal and opposite reaction". This statement means that in every interaction, there is a pair of forces acting on the two interacting objects. The size of the force on the first object equals the size of the force on the second object. The direction of the force on the first object is opposite to the direction of the force on the second object.

Example: A box is sitting on a table. The mass of the box is 4 kg. Because of the effects of gravity, the box is applying a force of 39.2 N on the table. The table does not break or shift under the force of the box. This implies that the table is applying a force of 39.2 N on the box. Note that the force that the table is applying to the box is in the opposite direction to the force that the box is applying to the table.

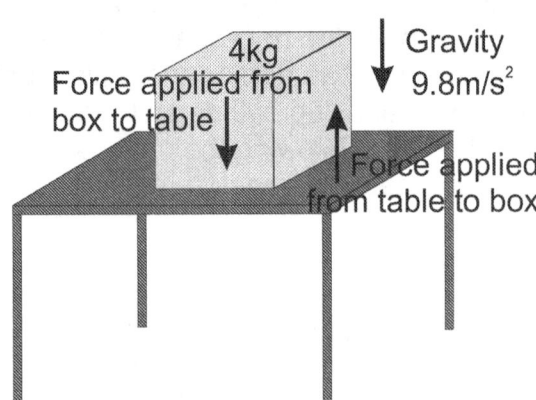

Here are a few more examples:

1. The propulsion/movement of fish through water: A fish uses its fins to push water backwards. The water pushes back on the fish. Because the force on the fish is unbalanced the fish moves forward.

2. The motion of car: A car's wheels push against the road and the road pushes back. Since the force of the road on the car is unbalanced the car moves forward.

3. Walking: When one pushes backwards on the foot with the muscles of the leg, the floor pushes back on the foot. If the forces of the leg on the foot and the floor on the foot are balanced, the foot will not move and the muscles of the body can move the other leg forward.

Skill 7.2 **Solving a variety of problems involving different types of forces in one and two dimensions**

Types of Forces:

Some of the common forces that act on a body are the following:

Gravity (See 7.4)

Normal force
When a body is pressed against a surface it experiences a reaction force that is perpendicular to the surface and in the direction away from the surface. For instance, an object resting on a table experiences an upward reaction force from the table that is equal and opposite to the force that the object exerts on the table. When the table is horizontal and no additional force is being applied to the object, the normal force is equal to the weight of the object.

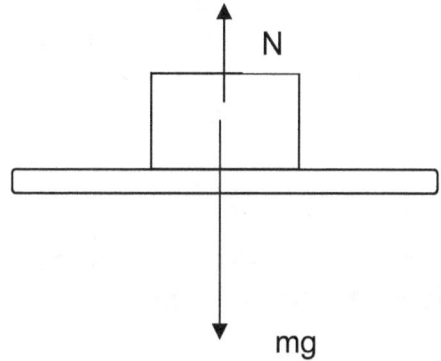

Friction
Friction is the force on a body that opposes its sliding over a surface. This force is due to the bonding between the two surfaces and is greater for rough surfaces. It acts in the direction opposite to the force attempting to move the object. When the object is at rest, the frictional force is known as **static friction**. The frictional force on an object in motion is known as **kinetic friction**.

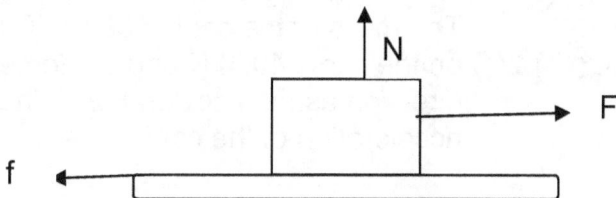

The frictional force is usually directly proportional to the normal force and can be calculated as $F_f = \mu F_n$ where μ is either the coefficient of static friction or kinetic friction depending on whether the object is at rest or in motion.

Tension/Compression

Tension is the force that acts in a rope, cable or rod that is attached to something and is being pulled. Tension acts along the cord. When a hand pulls a rope attached to a box, for instance, the tension T in the rope acts to pull the rope apart while it works on the box and the hand in the opposite direction as shown below:

Compression is the opposite of tension in that the force acts to shorten a rigid body instead of pulling it apart.

Net force

The forces that act on a body come from many different sources. Their effect on a body, however, is the same; a change in the state of motion of the body as given by Newton's laws of motion. Therefore, once we identify the magnitude and direction of each force acting on a body, we can combine the effect of all the forces together using vector addition and find the net force.

Newton's laws of motion can be used together or separately to analyze a variety of physical situations. Simple examples are provided below and the importance of each law is highlighted.

Problem:
A 10 kg object moves across a frictionless surface at a constant velocity of 5 m/s. How much force is necessary to maintain this speed?

Solution:
Both Newton's first and second laws can help us understand this problem. First, the first law tells us that this object will continue its state of uniform speed in a straight line (since there is no force acting upon it). Additionally, the second law tells that because there is no acceleration (velocity is constant), no force is required. Thus, zero force is necessary to maintain the speed of 5 m/s.

Problem:
A car is driving down a road at a constant speed. The mass of the car is 400 kg. The force acting on the car is 4000 N and the force is in the same direction as the acceleration. What is the acceleration of the car?

Solution: $F = ma \Rightarrow a = \dfrac{F}{m} \Rightarrow a = \dfrac{4000N}{400kg} =$
$a = 10 \, m/s^2$

Problem:
For the arrangement shown, find the force necessary to overcome the 500 N force pushing to the left and move the truck to the right with an acceleration of 5 m/s².

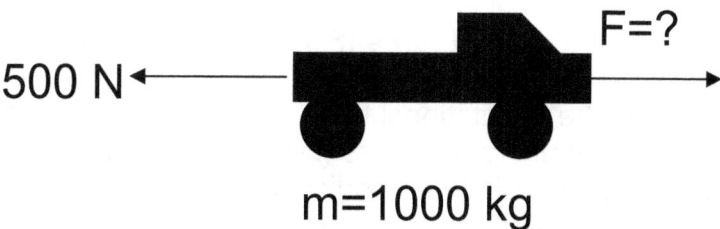

Solution:
The net force on the truck acting towards the right is F − 500N.
Using Newton's second law, F−500N = 1000kg × 5 m/s².
Solving for F, we get F = 5500 N.

Problem:
An astronaut with a mass of 95 kg stands on a space station with a mass of 20,000 kg. If the astronaut is exerting 40 N of force on the space station, what is the acceleration of the space station and the astronaut?

Solution
To find the acceleration of the space station, we can simply apply Newton's second law:

$$A_s = \frac{F}{m_s} = \frac{40N}{20000kg} = 0.002 \, m/s^2$$

To find the acceleration of the astronaut, we must first apply Newton's third law to determine that the space station exerts an opposite force of -40 N on the astronaut. Here the minus sign simply denotes that the force is directed in the opposite direction. We can then calculate the acceleration, again using Newton's second law:

$$A_a = \frac{F}{m_a} = \frac{-40N}{95kg} = -0.42 \, m/s^2$$

Frictional Forces:
In the real world, whenever an object moves its motion is opposed by a force known as friction. How strong the frictional force is depends on numerous factors such as the roughness of the surfaces (for two objects sliding against each other) or the viscosity of the liquid an object is moving through. Most problems involving the effect of friction on motion deal with sliding friction. This is the type of friction that makes it harder to push a box across cement than across a marble floor.

When you try and push an object from rest, you must overcome the maximum **static friction** force to get it to move. Once the object is in motion, you are working against **kinetic friction** which is smaller than the static friction force previously mentioned. Sliding friction is primarily dependent on two things, the **coefficient of friction (μ)** which is primarily dependent on roughness of the surfaces involved and the amount of force pushing the two surfaces together. This force is also known as the **normal force (F_n)**, the perpendicular force between two surfaces. When an object is resting on a flat surface, the normal force is pushing opposite to the gravitational force – straight up. When the object is resting on an incline, the normal force is less (because it is only opposing that portion of the gravitational force acting perpendicularly to the object) and its direction is perpendicular to the surface of incline but at an angle from the ground. Therefore, for an object experiencing no external action, the magnitude of the normal force is either equal to or less than the magnitude of the gravitational force (F_g) acting on it. The frictional force (F_f) acts perpendicularly to the normal force, opposing the direction of the object's motion.

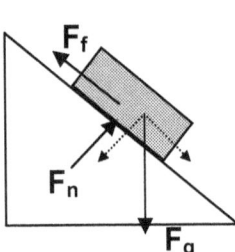

The frictional force is normally directly proportional to the normal force and, unless you are told otherwise, can be calculated as $F_f = \mu F_n$ where μ is either the coefficient of static friction or kinetic friction depending on whether the object starts at rest or in motion. In the first case, the problem is often stated as "how much force does it take to start an object moving" and the frictional force is given by $F_f > \mu_s F_n$ where μ_s is the coefficient of static friction. When questions are of the form "what is the magnitude of the frictional force opposing the motion of this object," the frictional force is given by $F_f = \mu_k F_n$ where μ_k is the coefficient of kinetic friction.

There are several important things to remember when solving problems about friction:

1. The frictional force acts in opposition to the direction of motion.

2. The frictional force is proportional, and acts perpendicular to, the normal force.

3. The normal force is perpendicular to the surface the object is lying on. If there is a force pushing the object against the surface, it will increase the normal force.

Problem:
A woman is pushing an 800N box across the floor. She pushes with a force of 1000 N. The coefficient of kinetic friction is 0.50.

If the box is already moving, what is the force of friction acting on the box?

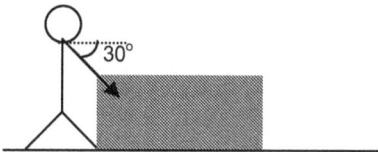

Solution:

First it is necessary to solve for the normal force.

F_n = 800N + 1000N (sin 30°) = 1300N

Then, since $F_f = \mu F_n$ = 0.5*1300=650N

Skill 7.3 Analyzing the vector nature of force

Vectors are used in representing any quantity that has both magnitude and direction. We must know not only that the force of gravity has a magnitude of 9.8 m/s² times the mass of an object, but that it is directed toward the center of the earth.

When we wish to analyze a physical situation involving vector quantities such as force, the first step is typically the creation of a diagram. Free body diagrams are simple sketches that show all the objects and forces in a given physical situation. This makes them very useful for understanding and solving physical problems. The objects involved are drawn and arrows are used to represent the vector quantities which are labeled appropriately. In precise diagrams drawn to scale on graph paper, the magnitude of each force is indicated by the length of the arrow and the exact angles between the vectors will be depicted. Otherwise, angles and magnitudes may simply be noted on the diagram.

Problem:
Find the net force on a 5 Kg box sliding down an inclined surface at an angle of $30°$ with the horizontal if the coefficient of friction of the surface is 0.5.

Solution:
There are three forces acting on the box, gravity, the normal force and the frictional force. We can resolve these forces along the inclined plane and perpendicular to the plane and find the net force in each direction.

Perpendicular to the plane:
The component of the gravitational force perpendicular to the plane = mgcos30 = 5 x 9.8 x 0.87 = 42.6N

The normal force acting on the box is equal and opposite to the perpendicular component of the gravitational force. Thus the net force on the box perpendicular to the plane is zero.

Along the inclined plane:

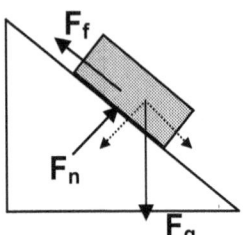

The component of the gravitational force down the plane
=mgsin30 = 5 x 9.8 x 0.5 = 24.5N
The force of friction up the plane = μ F_n = 0.5 x 42.6 = 21.3N
Thus the net force on the box acts down the plane and is equal to 24.5 – 21.3 = 3.2N

Skill 7.4 Determining methods for measuring force and differentiating between mass and weight

The gravitational force pulls a body towards the center of the earth, i.e. downwards, and is also called the weight of the body.

It is given by, $W = mg$, where m is the mass of the body and $g = 9.81$ m/s² is the acceleration due to gravity.

It is important to recognize that an object has the same mass on Earth as it does on Jupiter. However, because Jupiter has greater mass than the Earth, the value of "g" (the acceleration of due to gravity) on Jupiter will be greater than "g" on the Earth. Therefore, an object's weight (the product of mass and acceleration due to gravity) will be greater on Jupiter than on Earth.

The same principle holds for a person's mass and weight on the moon. The moon is much less massive than the earth and therefore has a smaller value for "g". Therefore, you would weigh approximately six times less, but have the same mass regardless of being on the moon!

Designing an experiment to accurately measure a force directly can be extremely challenging and indirect methods are often required. For example, to measure the force on a moving object, it may be simpler to measure the position of the object with time, differentiate the results twice to get the acceleration and then multiply by the mass of the object in accordance with $F = ma$. Thus, force measurement experiments may often be more complicated than simply using a mechanical scale delineated in Newtons.

Indirect force measurement through fields or potentials

Forces may be measured indirectly through a field or potential measurement. Since it is impossible to attach a tiny spring scale to an electron in order to measure the force of an electric field upon it, a measurement of the electric field or electric potential can be used instead. Such measurements may be taken by way of a field meter or multimeter, providing information about the field or potential from which the force on an electron (or other object) can be calculated.

Indirect force measurements through material characteristics

Properties of various materials can also be used to indirectly measure a force. The piezoelectric effect is one example that results from the ability of certain materials, such as some crystals and ceramics, to produce an electric potential when they are stressed mechanically. Thus, these natural "pressure sensors" can be used for force measurements through measurement of the voltage produced in a given situation. One particular application of piezoelectric materials is in equipment for fine weight measurements that can be used to measure the gravitational force on small or low-mass objects.

The temperature and volume of a confined gas may also be used to measure force by using, for example, a piston. Force applied to the piston causes an increase in the pressure of the gas and, by the ideal gas law ($PV = nRT$), a corresponding change in the volume and temperature of the gas. The pressure can be calculated from these parameters (or measured directly), thus allowing the calculation of the force on the piston.

Stages of experimental design

Experimental apparatuses for measurement of forces, therefore, involve several aspects. First, the transducer, which may be more or less a direct measurement of the force, acts as a converter of force into some intelligible signal. In some cases, as with the examples of the piston or the piezoelectric potential, the magnitude of the force must be calculated from the signal (whether electric or otherwise) through a theoretical relationship determined by the characteristics of the material or materials used. Second, for the more indirect force measurements, computer equipment may be required for calculating the force based on other parameters in the experiment. In the example of the moving object, it is most convenient to use a computer for performing the differentials, although, for sufficiently small data sets, this can be done "by hand." Regardless of the complexity, an algorithm for calculating the force values must be followed, and this is as much part of the experimental apparatus as the transducer equipment. In simple cases, however, such as spring scales, the force can be measured by simply looking at the reading.

Skill 7.5 Applying the law of universal gravitation and Kepler's laws in a variety of situations

Newton's universal law of gravitation states that any two objects experience a force between them as the result of their masses. Specifically, the force between two masses m_1 and m_2 can be summarized as

$$F = G\frac{m_1 m_2}{r^2}$$

where G is the gravitational constant ($G = 6.672 \times 10^{-11} \, Nm^2/kg^2$), and r is the distance between the two objects.

The weight of an object is the result of the gravitational force of the earth acting on its mass. The acceleration due to Earth's gravity on an object is 9.81 m/s². Since force equals mass * acceleration, the magnitude of the gravitational force created by the earth on an object is

$$F_{Gravity} = m_{object} \cdot 9.81 \, m/s^2$$

Important things to remember:

1. The gravitational force is proportional to the masses of the two objects, but *inversely* proportional to the *square of the distance* between the two objects.

2. When calculating the effects of the acceleration due to gravity for an object above the earth's surface, the distance above the surface is ignored because it is inconsequential compared to the radius of the earth. The constant figure of 9.81 m/s² is used instead.

Problem: Two identical 4 kg balls are floating in space, 2 meters apart. What is the magnitude of the gravitational force they exert on each other?

Solution:

$$F = G\frac{m_1 m_2}{r^2} = G\frac{4 \times 4}{2^2} = 4G = 2.67 \times 10^{-10} \, N$$

Planetary Motion: Johannes Kepler was a German mathematician who studied the astronomical observations made by Tyco Brahe. He derived the following three laws of planetary motion. Kepler's laws also predict the motion of comets.

First Law
This law describes the shape of planetary orbits. Specifically, the orbit of a planet is an ellipse that has the sun at one of the foci. Such an orbit looks like this:

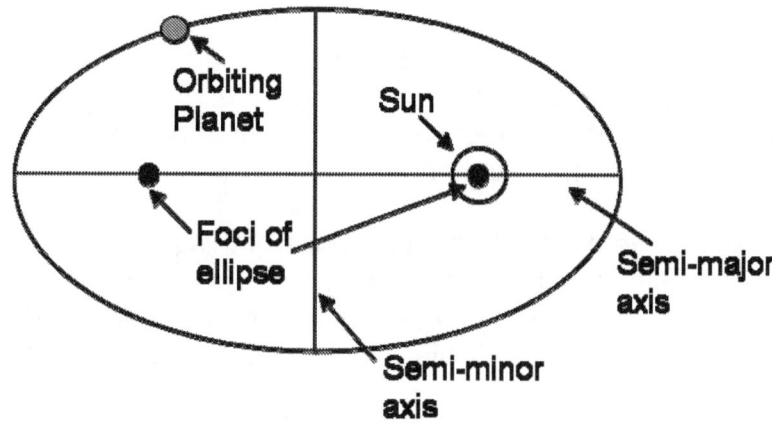

To analyze this situation mathematically, remember that the semi-major axis is denoted a, the semi-minor axis denoted b, and the general equation for an ellipse in polar coordinates is:

$$r = \frac{l}{1 + e \cos \theta}$$

Where r=radial coordinate
θ=angular coordinate
l= semi-latus rectum ($l=b^2/a$)
e=eccentricity (for an ellipse, $e = \sqrt{1 - \frac{b^2}{a^2}}$)

Thus, we can also determine the planet's maximum and minimum distance from the sun.

The point at which the planet is closest to the sun is known as the perihelion and occurs when θ=0:
$$r_{min} = \frac{l}{1+e}$$

The point at which the planet is farthest from the sun is known as the aphelion and occurs when θ=180°:

$$r_{max} = \frac{l}{1-e}$$

Second Law

The second law pertains to the relative speed of a planet as it orbits. This law says that a line joining the planet and the Sun sweeps out equal areas in equal intervals of time. In the diagram below, the two shaded areas demonstrate equal areas.

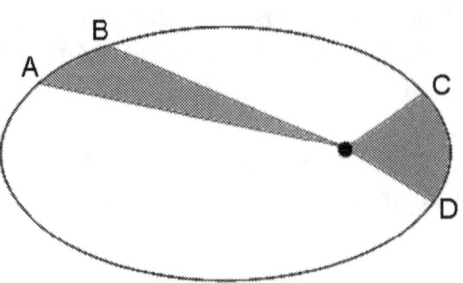

By Kepler's second law, we know that the planet will take the same amount of time to move between points A and B and between points C and D. Note that this means that the speed of the planet is inversely proportional to its distance from the sun (i.e., the plant moves fastest when it is closest to the sun). You can view an animation of this changing speed here: http://home.cvc.org/science/kepler.gif

Kepler was only able to demonstrate the existence of this phenomenon but we now know that it is an effect of the Sun's gravity. The gravity of the Sun pulls the planet toward it thereby accelerating the planet as it nears. Using the first two laws together, Kepler was able to calculate a planet's position from the time elapsed since the perihelion.

Third Law

The third law is also known as the harmonic law and it relates the size of a planet's orbital to the time needed to complete it. It states that the square of a planet's period is proportional to the cube of its mean distances from the Sun (this mean distance can be shown to be equal to the semi-major axis). So, we can state the third law as:

$$P^2 \propto a^3$$

where P = planet's orbital period (length of time needed to complete one orbit)
a = semi-major axis of orbit

Furthermore, for two planets A and B: $P_A^2 / P_B^2 = a_A^3 / a_B^3$

The units for period and semi-major axis have been defined such that $P^2 a^{-3}=1$ for all planets in our solar system. These units are sidereal years (yr) and astronomical units (AU). Sample values are given in the table below. Note that in each case $P^2 \sim a^3$

Planet	P (yr)	a (AU)	P^2	a^3
Venus	0.62	0.72	0.39	0.37
Earth	1.0	1.0	1.0	1.0
Jupiter	11.9	5.20	142	141

COMPETENCY 8.0 UNDERSTAND CONSERVATION OF ENERGY AND CONSERVATION OF MOMENTUM

Skill 8.1 Includes applying the concepts of work, energy, and power in a variety of situations

In physics, work is defined as force times distance $W = F \cdot s$. Work is a scalar quantity, it does not have direction, and it is usually measured in Joules ($N \cdot m$). It is important to remember, when doing calculations about work, that the only part of the force that contributes to the work is the part that acts in the direction of the displacement. Therefore, sometimes the equation is written as $W = F \cdot s \cos\theta$, where θ is the angle between the force and the displacement.

Problem:
A man uses 6N of force to pull a 10kg block, as shown below, over a distance of 3 m. How much work did he do?

Solution:

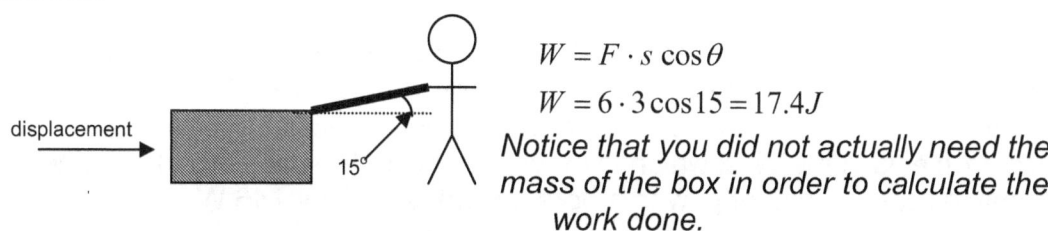

$W = F \cdot s \cos\theta$

$W = 6 \cdot 3 \cos 15 = 17.4 J$

Notice that you did not actually need the mass of the box in order to calculate the work done.

Energy is also defined, in relation to work, as the ability of an object to do work. As such, it is measured in the same units as work, usually Joules. Most problems relating work to energy are looking at two specific kinds of energy. The first, kinetic energy, is the energy of motion. The heavier an object is and the faster it is going, the more energy it has resulting in a greater capacity for work. The equation for kinetic energy is: $KE = \frac{1}{2}mv^2$.

Problem:
A 1500 kg car is moving at 60m/s down a highway when it crashes into a 3000kg truck. In the moment before impact, how much kinetic energy does the car have?

Solution:
$KE = \frac{1}{2}mv^2 = \frac{1}{2} \cdot 1500 \cdot 60^2 = 2.7 \times 10^6 J$

The other form of energy frequently discussed in relationship to work is gravitational potential energy, or potential energy, the energy of position.

Potential energy is calculated as $PE = mgh$ where h is the distance the object is capable of falling.

Problem:
Which has more potential energy, a 2 kg box held 5 m above the ground or a 10 kg box held 1 m above the ground?

Solution:

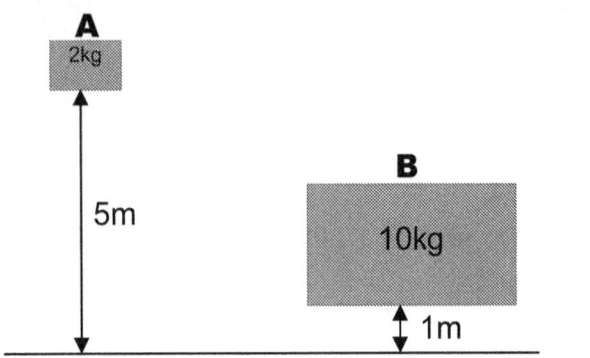

$$PE_A = mgh = 2 \cdot g \cdot 5 = 10g$$
$$PE_B = mgh = 10 \cdot g \cdot 1 = 10g$$
$$PE_A = PE_B$$

The power expended by a system can be defined as either the rate at which work is done by the system or the rate at which energy is transferred from it. There are many different measurements for power, but the one most commonly seen in physics problems is the Watt which is measured in Joules per second. Another commonly discussed unit of power is horsepower, and 1hp=746 W.

The **average power** of a system is defined as the work done by the system divided by the total change in time:

$\overline{P} = \dfrac{W}{\Delta t}$ ⇒ Where $\overline{P}$ = average power, W = work and Δt = change in time

The average power can also be written in terms of energy transfer ⇒ $\overline{P} = \dfrac{\Delta E}{\Delta t}$

and used the same way that the equation for work is used.

Problem:
A woman standing in her 4th story apartment raises a 10kg box of groceries from the ground using a rope. She is pulling at a constant rate, and it takes her 5 seconds to raise the box one meter. How much power is she using to raise the box?

Solution:

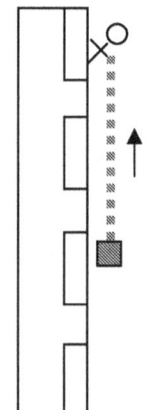

$$P = W/t$$
$$P = \frac{F \cdot s}{t} = \frac{mgh}{t} = \frac{10*9.8*1}{5} = 19.6W$$

Notice that because she is pulling at a constant rate, you don't need to know the actual distance she has raised the box. 2 meters in 10 seconds would give you the same result as 5 meters in 25 seconds.

Instantaneous power is the power measured or calculated in an instant of time. Since instantaneous power is the rate of work done when Δt is approaching 0s, the power is then written in derivate form:

$P = \frac{dW}{dt} \Rightarrow$ Where P = average power, dW = work and dt = change in time.

Since $W = Fs\cos\phi$, for a constant force the above equation can be written as:

$$P = \frac{dW}{dt} \Rightarrow P = \frac{d(Fs\cos\phi)}{dt} \Rightarrow P = \frac{(F\cos\phi)ds}{dt} \Rightarrow P = F\cos\phi\left(\frac{ds}{dt}\right) \Rightarrow P = Fv\cos\phi$$

where v is the velocity of the object.

Skill 8.2   Analyzing the kinetic and potential energy of various systems

According to the concept of conservation of energy, the energy in an isolated system remains the same although it may change in form. For instance, potential energy can become kinetic energy and kinetic energy, depending on the system, can become thermal or heat energy. Solving energy conservation problems depends on knowing the types of energy one is dealing with in a particular situation and assuming that the sum of all the different types of energy remains constant. Below we will discuss several different examples.

Example:

A rollercoaster at the top of a hill has a certain potential energy that will allow it to travel down the track at a speed based on its potential energy and friction with the track itself. At the bottom of the hill, when it has reached a stop, its potential energy is zero and all of the energy has been transferred from potential energy to kinetic energy (movement) and thermal energy (heat derived from friction). The equation below describes the relationship between potential energy and other forms of energy in this case:

Potential Energy = Kinetic energy(movement) + heat energy(friction)

Problem:

A skier travels down a ski slope with negligible friction. He begins at 100 meters in height, drops to a much lower level and ends at 90 meters in height. What is the skier's velocity at the 90 meter height?

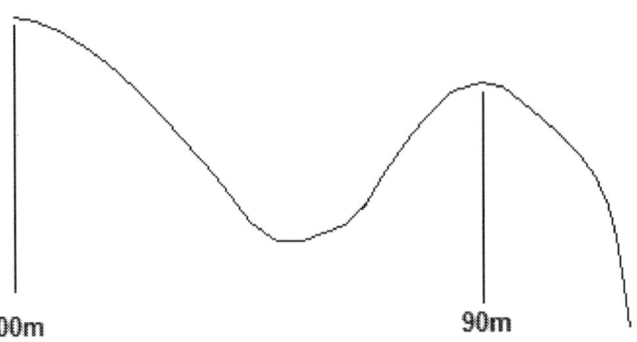

The skier's initial energy is only potential energy and is given by mgh = 100mg where m is the mass of the skier.
The skier's final energy is the sum of his potential and kinetic energies and is given by mgh + 1/2mv^2 = 90mg + 1/2mv^2 where v is the skier's velocity.
Using the principle of conservation of energy we know that the initial and final energies must be equal. Hence, 100mg = 90mg + 1/2mv^2

Thus 1/2mv^2 = 10mg; 1/2v^2 = 10g; v = 14m/s since (g=9.81m/s^2)

Problem:

d. A pebble weighing 10 grams is placed in a massless frictionless sling shot, with a spring constant of 200N/m, that is stretched back 0.5 meters. What is the total energy of the system before the pebble is released? What is the final height of the pebble if it is shot straight up and the effects of air resistance are negligible?

Solution:

If the initial height of the pebble is h = 0, total energy is given by
E = ½ mv^2 + mgh + ½ kx^2 = 0 + 0 + 0.5(200)(0.5)^2 = 25 Joules

At its final height, the velocity of the pebble will be zero. Since
E = ½ mv^2 + mgh + ½ kx^2, from the principle of conservation of energy
25 Joules = 0 + 0.010kg (9.81)h + 0 and h = 255 m

Skill 8.3 Applying the principles of conservation of energy and conservation of linear momentum to situations, including elastic and inelastic collisions

The law of **conservation of linear momentum** states that the total momentum of an *isolated system* (not affected by external forces and not having internal dissipative forces) always remains the same. For instance, in any collision between two objects in an isolated system, the total momentum of the two objects after the collision will be the same as the total momentum of the two objects before the collision. In other words, any momentum lost by one of the objects is gained by the other.

A collision may be **elastic** or **inelastic**. In a totally elastic collision, the kinetic energy is conserved along with the momentum. In a totally inelastic collision, on the other hand, the kinetic energy associated with the center of mass remains unchanged but the kinetic energy relative to the center of mass is lost. An example of a totally inelastic collision is one in which the bodies stick to each other and move together after the collision. Most collisions are neither perfectly elastic nor perfectly inelastic and only a portion of the kinetic energy relative to the center of mass is lost.

Imagine two carts rolling towards each other as in the diagram below

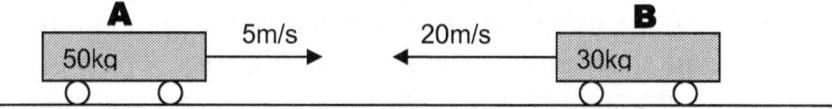

Before the collision, cart **A** has 250 kg m/s of momentum, and cart **B** has –600 kg m/s of momentum. In other words, the system has a total momentum of –350 kg m/s of momentum.

After the inelastic collision, the two cards stick to each other, and continue moving. How do we determine how fast, and in what direction, they go?

We know that the new mass of the cart is 80kg, and that the total momentum of the system is –350 kg m/s. Therefore, the velocity of the two carts stuck together must be $\frac{-350}{80} = -4.375 \, m/s$

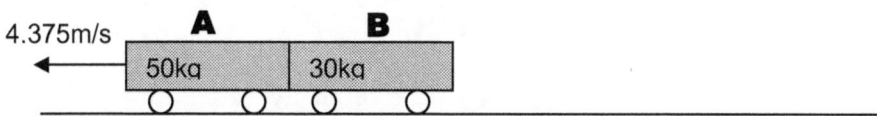

Conservation of momentum works the same way in two dimensions, the only change is that you need to use vector math to determine the total momentum and any changes, instead of simple addition.

Imagine a pool table like the one below. Both balls are 0.5 kg in mass.

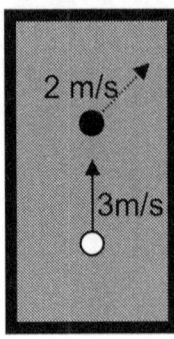

Before the collision, the white ball is moving with the velocity indicated by the solid line and the black ball is at rest.

After the collision the black ball is moving with the velocity indicated by the dashed line (a 135° angle from the direction of the white ball).

With what speed, and in what direction, is the white ball moving after the collision?

$$p_{white/before} = .5 \cdot (0,3) = (0,1.5) \quad p_{black/before} = 0 \quad p_{total/before} = (0,1.5)$$

$$p_{black/after} = .5 \cdot (2\cos 45, 2\sin 45) = (0.71, 0.71)$$

$$p_{white/after} = (-0.71, 0.79)$$

i.e. the white ball has a velocity of $v = \sqrt{(-.71)^2 + (0.79)^2} = 1.06 m/s$

and is moving at an angle of $\theta = \tan^{-1}\left(\dfrac{0.79}{-0.71}\right) = -48°$ from the horizontal

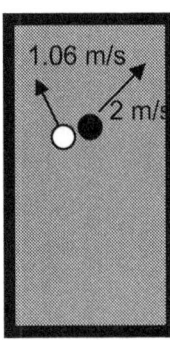

The **impulse-momentum theorem** states that any impulse acting on a system changes the momentum of that system. When considering the impulse-momentum theorem, there are several factors that need to be taken into account. The first factor is that momentum is a vector quantity $p = m \cdot v$. It has both magnitude and direction. Therefore, any action that causes either the speed or the direction of an object to change causes a change in its momentum. An impulse is defined as a force acting over a period of time (integral of force over time), and any impulse acting on the system is equivalent to a change in its momentum, as you can see from the equations below:

$$F = m \cdot a \rightarrow F = m \cdot \frac{\Delta v}{t} \rightarrow F \cdot t = m \cdot \Delta v$$

i.e. Forces acting over time cause a change in momentum.

Sample Problems:

1. A 1 kg ball is rolled towards a wall at 4 m/s. It hits the wall, and bounces back off the wall at 3 m/s.

What is the change in velocity?

The velocity goes from +4m/s to –3m/s, a net change of -7m/s.

At what point does the impulse occur?

The impulse occurs when the ball hits the wall.

2. A 30kg woman is in a car accident. She was driving at 50m/s when she had to hit the brakes to avoid hitting the car in front of her.

The automatic tensioning device in her seatbelt slows her down to a stop over a period of one half second. How much force does it apply?

$$F = m \cdot \frac{\Delta v}{t} \rightarrow F = 30 \cdot \frac{50}{.5} = 3000 N$$

If she hadn't been wearing a seatbelt, the windshield would have stopped her in .001 seconds. How much force would have been applied there?

$$F = m \cdot \frac{\Delta v}{t} \rightarrow F = 30 \cdot \frac{50}{.001} = 1500000 N$$

TEACHER CERTIFICATION STUDY GUIDE

COMPETENCY 9.0 UNDERSTAND TORQUE, STATIC EQUILIBRIUM, AND ROTATIONAL DYNAMICS

Skill 9.1 Analyzing the forces and torques acting in a given situation

Linear motion is measured in rectangular coordinates. Rotational motion is measured differently, in terms of the angle of displacement. There are three common ways to measure rotational displacement; degrees, revolutions, and radians. Degrees and revolutions have an easy to understand relationship, one revolution is 360°. Radians are slightly less well known and are defined as

$$\frac{arc\ length}{radius}.$$ Therefore 360°=2π radians and 1 radian = 57.3°.

The major concepts of linear motion are duplicated in rotational motion with linear displacement replaced by **angular displacement**.

Angular velocity ω = rate of change of angular displacement.
Angular acceleration α = rate of change of angular velocity.

Torque is rotational motion about an axis. It is defined as $\tau = L \times F$, where L is the lever arm. The length of the lever arm is calculated by measuring the perpendicular line drawn from the line of force to the axis of rotation.

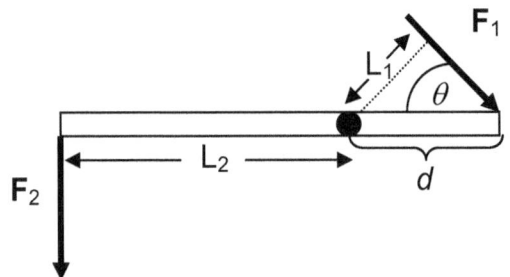

By convention, torques that act in a clockwise direction are considered negative and those in a counterclockwise direction are considered positive. In order for an object to be in equilibrium, the sum of the torques acting on it must be zero.

The equation that would put the above figure in equilibrium is $F_1 L_1 = -F_2 L_2$ (please note that in this case L$_1$=$d\sin\theta$).

Examples:
Some children are playing with the spinner below, when one young boy decides to pull on the spinner arrow in the direction indicated by **F$_1$**. How much torque does he apply to the spinner arrow?

$\tau = L \times F$, but L=0 because the line of force goes directly through the axis of rotation (i.e. the perpendicular distance from the line of force to the pivot point is 0). Therefore the boy applies no torque to the spinner.

TEACHER CERTIFICATION STUDY GUIDE

Problem: In the system diagrammed below, find out what the magnitude of F_1 must be in order to keep the system in equilibrium.

Solution: For an object to be in equilibrium the forces acting on it must be balanced. This applies to linear as well as rotational forces known as moments or torques.

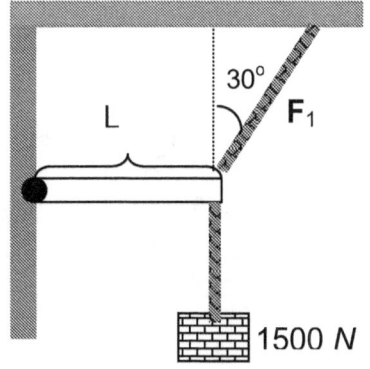

$$F_1 L = -F_2 L$$
$$F_{1x} L + F_{1y} L = -(-1500L) \ldots \text{but } F_{1x} = 0 *$$
$$F_{1y} L = 1500L$$
$$F_1 \cos 30 L = 1500L$$
$$(0.866) F_1 = 1500$$
$$F_1 = 1732 N$$

*the effect of $F_{1x}=0$ because that portion of the force goes right through the pivot and causes no torque.

Things to remember:

1. When writing the equation for a body at equilibrium, the point chosen for the axis of rotation is arbitrary.

2. The center of gravity is the point in an object where its weight can be considered to act for the purpose of calculating torque.

3. The lever arm, or moment arm, of a force is calculated as the perpendicular distance from the line of force to the pivot point/axis of rotation.

4. Counterclockwise torques are considered positive while clockwise torques are considered negative.

Skill 9.2 Applying the concepts of force, torque, and energy to analyze the operation of simple devices

Work is defined as the product of force and distance and **power** as the rate at which work is done. A simple machine, consisting of a single mechanical device, is often used to "make work easier". A machine can be considered in terms of input force and output force. Additionally, when considering the distances over which those forces are applied there is input and output work.

In the ideal case of a frictionless process, the input work equals the output work. However, simple machines can effectively make tasks easier is by requiring less force. For example, some work is done by a force F_1 over a distance D_1. The same work can be accomplished by a lesser force $F_2 < F_1$ over a greater distance $D_2 > D_1$.

$$\text{Input Work} = \text{Output Work}$$
$$(\text{Input Force})(\text{Distance}) = (\text{Output Force})(\text{Distance})$$
$$F_1 D_1 = F_2 D_2$$

Reduced force over a longer distance is the concept behind an inclined plane. Other simple machines operate by different principles that "make work easier" such as transferring a force from one place to another, changing the direction of a force, increasing the magnitude of a force, or increasing the distance or speed of a force. The six types of simple machines:

- Inclined Plane
- Lever
- Wheel and Axle
- Pulley
- Wedge
- Screw

The efficiency of a machine is percentage of input work that converted to usable output work or the ratio of work output to work input.

$$\text{Efficiency} = \frac{\text{Work Output}}{\text{Work Input}}$$

As we know from the second law of thermodynamics, the efficiency of a machine can never equal 100%. Typically, the energy not used for work is "lost" as heat.

TEACHER CERTIFICATION STUDY GUIDE

Skill 9.3 Applying the conservation of angular momentum

The major concepts of linear motion are duplicated in rotational motion with linear displacement replaced by the angle of displacement.

Angular velocity ω = angular displacement / time

Also, the linear velocity v of a rolling object can be written as $v = r\omega$.

One important difference in the equations relates to the use of mass in rotational systems. In rotational problems, not only is the mass of an object important but also its location. In order to include the spatial distribution of the mass of the object, a term called moment of inertia is used, $I = m_1 r_1^2 + m_2 r_2^2 + \cdots + m_n r_n^2$. The moment of inertia is always defined with respect to a particular axis of rotation.

Example: If the radius of the wheel on the left is 0.75m, what is its moment of inertia about an axis running through its center perpendicular to the plane of the wheel?

$$I = 3 \cdot 0.75^2 + 3 \cdot 0.75^2 + 3 \cdot 0.75^2 + 3 \cdot 0.75^2 = 6.75$$

Note: $I_{Sphere} = \frac{2}{5}mr^2$, $I_{Hoop/Ring} = mr^2$, $I_{disk} = \frac{1}{2}mr^2$

Similar to the conservation of linear momentum (Skill 8.3), unless a net torque acts on a system, the angular momentum remains constant in both magnitude and direction. This can be used to solve many different types of problems including ones involving satellite motion.

Example: A planet of mass m is circling a star in an orbit like the one below. If its velocity at point A is 60,000m/s, and $r_B = 8\, r_A$, what is its velocity at point B?

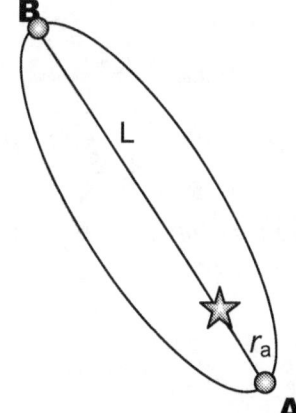

$$I_B \omega_B = I_A \omega_A$$
$$mr_B^2 \omega_B = mr_A^2 \omega_A$$
$$r_B^2 \omega_B = r_A^2 \omega_A$$
$$r_B^2 \frac{v_B}{r_B} = r_A^2 \frac{v_A}{r_A}$$
$$r_B v_B = r_A v_A$$
$$8 r_A v_B = r_A v_A$$
$$v_B = \frac{v_A}{8} = 7500 m/s$$

PHYSICS

Skill 9.4 Analyzing the motion of a rigid body in terms of moment of inertia, rotational kinetic energy, and angular momentum

The rotational analog of Newton's second law of motion is given in terms of torque τ, moment of inertia I, and angular acceleration α:

$$\tau = I\alpha$$

where the torque τ is the rotational force on the body. In simple terms, the torque τ produced by a force F acting at a distance r from the point of rotation is given by the product of r and the component of the force that is perpendicular to the line joining the point of rotation to the point of action of the force.

Angular momentum (L), and rotational kinetic energy (KE_r), are therefore defined as follows: $L = I\omega, \quad KE_r = \frac{1}{2}I\omega^2$

As with all systems, energy is conserved unless the system is acted on by an external force. This can be used to solve problem such as the one below.

Example:

A uniform ball of radius *r* and mass *m* starts from rest and rolls down a frictionless incline of height *h*. When the ball reaches the ground, how fast is it going?

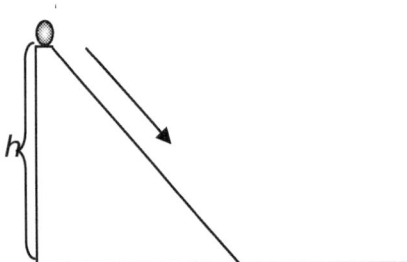

$$PE_{initial} + KE_{rotational/initial} + KE_{linear/initial} = PE_{final} + KE_{rotational/final} + KE_{linear/final}$$

$$mgh + 0 + 0 = 0 + \frac{1}{2}I\omega_{final}^2 + \frac{1}{2}mv_{final}^2 \rightarrow mgh = \frac{1}{2} \cdot \frac{2}{5}mr^2\omega_{final}^2 + \frac{1}{2}mv_{final}^2$$

$$mgh = \frac{1}{5}mr^2(\frac{v_{final}}{r})^2 + \frac{1}{2}mv_{final}^2 \rightarrow mgh = \frac{1}{5}mv_{final}^2 + \frac{1}{2}mv_{final}^2$$

$$gh = \frac{7}{10}v_{final}^2 \rightarrow v_{final} = \sqrt{\frac{10}{7}gh}$$

COMPETENCY 10.0 UNDERSTAND THE CHARACTERISTICS OF OSCILLATORY MOTION

Skill 10.1 Analyzing models of simple harmonic motion

Simple Pendulum

The simple pendulum model provides a reasonably accurate representation of pendulum motion, especially in the case of small angular amplitude.

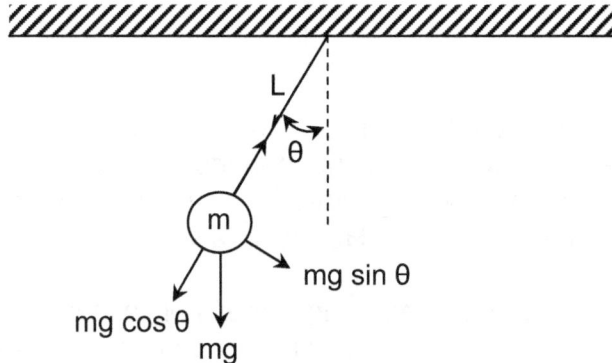

The restoring force F in pendulum motion is expressed as a component of the gravitational force mg perpendicular to the length of the string and is given by

$$F = -mg\sin\theta$$

The negative sign results from the force having a direction opposite to the displacement. The tension T on the string and the portion of the gravitational force in the opposite direction of T cancel one another.

In the case of small θ, sin θ is approximately equal to θ. The arc length traveled by the pendulum, s, is equal to the product of the length L of the string and the angle θ. Thus, the following expression can be derived.

$$F \approx -mg\theta = -mg\frac{s}{L} = -\frac{mg}{L}s$$

In the above equation, the force is shown to be of the same form as the linear harmonic oscillator, having, in this case, a "spring constant" of mg/L. As a result, the expressions found for the displacement, velocity and acceleration in the case of the linear oscillator can also be used here (in the case of small θ), where the frequency ω is replaced as follows.

$$\omega = \sqrt{\frac{g}{L}}$$

Mass on a Spring

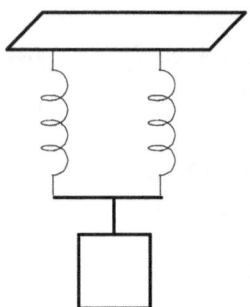

The above diagram is an example of a **Hookean system** (a spring, wire, rod, etc.) where the spring returns to its original configuration after being displaced and then released. When the spring is stretched a distance x, the restoring force exerted by the spring is expressed by Hooke's Law

$$F = -kx$$

The minus sign indicates that the restoring force is always opposite in direction to the displacement. The variable k is the spring constant and measures the stiffness of the spring in N/m.

The period of simple harmonic motion for a Hookean spring system is dependent upon the mass (m) of the spring and the stiffness of the spring (k) and is given by

$$T = \frac{1}{f} = \frac{1}{\frac{\omega}{2\pi}} = \frac{2\pi}{\omega} = \frac{2\pi}{\sqrt{\frac{k}{m}}} = 2\pi\sqrt{\frac{m}{k}}$$

Problem: Each spring in the above diagram has a stiffness of $k = 20 N/m$. The mass of the object connected to the spring is 2 kg. Ignoring friction forces, find the period of motion.

Solution: Utilizing Hooke's Law, the net restoring force on the spring would be

$$F = -(20N/m)x - (20N/m)x = -(40N/m)x$$

Comparison with $F = -kx$ shows the equivalent k to be 40 N/m.
Using the above formula for our problem, we have

$$T = 2\pi\sqrt{\frac{m}{k}} = 2\pi\sqrt{\frac{2kg}{40N/m}} = 2(3.14)\sqrt{0.05} = 1.4s$$

Skill 10.2 Recognizing the relationship between the simple harmonic oscillator and uniform circular motion

Simple harmonic (sinusoidal) motion involves a cyclical exchange of kinetic energy and potential energy as observed in a simple pendulum or a mass on a spring. The relationships among the various parameters of a system displaying simple harmonic motion depends on the type of system being examined. Once the displacement is known, the velocity and acceleration of the object undergoing harmonic motion can be calculated by calculating the first derivative with respect to time (for velocity) or the second derivative with respect to time (for acceleration).

The examples of the mass on a spring and the simple pendulum represent harmonic motion in one dimension and two dimensions, respectively. It can be shown, however, that the pendulum acts just like the mass on a spring when the displacement angle is small. Given the frequency of oscillation f for a mass on a spring (a linear oscillator), along with the concomitant period T = 1/f, the displacement of the mass undergoing harmonic (sinusoidal) motion can be written as follows.

$$x(t) = x_{max} \cos(2\pi f t + \phi) = x_{max} \cos(\omega t + \phi)$$

Alternatively, 2πf can be written as the angular frequency ω. The coefficient x_{max} is the maximum displacement of the mass, and the term Φ is a phase constant that determines the position of the mass at time t = 0. If x(t) is differentiated once with respect to time, the velocity of the mass is revealed.

$$v(t) = \frac{\partial x(t)}{\partial t} = -\omega x_{max} \sin(\omega t + \phi)$$

Comparing the expressions for displacement and velocity, we can see that the velocity is maximized when the displacement is zero (all kinetic energy), and the displacement is maximized when the velocity is zero (all potential energy); that is, the displacement and velocity are 90° out of phase. The acceleration can be calculated by differentiating v(t) with respect to time.

$$a(t) = \frac{\partial v(t)}{\partial t} = -\omega^2 x_{max} \cos(\omega t + \phi) = -\omega^2 x(t)$$

The acceleration, as shown above, is in phase with the displacement. Applying Newton's second law of motion leads to Hooke's law, which relates the restoring force on the mass to the displacement x(t). $F = ma = -m\omega^2 x(t)$

This equation may be expressed in terms of the so-called spring constant k, which is defined as $m\omega^2$. $F = ma = -k x(t)$

PHYSICS

Skill 10.3 Applying the law of conservation of energy to oscillating systems

Though we know total energy is always conserved, there are many physical everyday examples in which we can see how kinetic energy is lost when it is converted to potential energy and vice versa. Recall that kinetic energy is energy possessed by an object due to motion and potential energy is that stored by an object due to position associated with a force.

Total energy must always be conserved, but in an oscillating system that energy is alternately in the form of kinetic and potential energy. Imagine, for instance, a mass connected to a spring. As the spring is compressed, elastic potential energy is stored within the spring. Specifically, when the spring is maximally compressed it also has maximum potential energy and minimum kinetic energy (zero kinetic energy). When the spring is released, this potential energy converts back into kinetic energy as the mass travels outward. This conversion back and forth happens continually as the spring oscillates.

For a mass on a spring with spring constant k, amplitude of oscillation A, angular frequency ω, time t, and phase Φ:

$$\text{Kinetic energy} = \tfrac{1}{2} k A^2 \sin^2(\omega_0 t + \Phi)$$

$$\text{Potential energy} = \tfrac{1}{2} k A^2 \cos^2(\omega_0 t + \Phi)$$

$$\text{Total energy} = \tfrac{1}{2} k A^2$$

We know that sine and cosine are 90° out of phase. Thus, this is a mathematical statement that when potential energy is maximal, kinetic energy is minimum. Further, kinetic energy plus potential energy is always equal to a constant, that is, total energy in the system is always conserved.

Skill 10.4 Recognizing the effects of damping

Harmonic oscillators were introduced in **Skill 10.1,** but no discussion of how these oscillators actually cease moving and return to equilibrium was presented. From real world experience, however, we know that over time the oscillations reduce in amplitude and eventually the system comes to rest. This effect is known as damping.

The damping force is directionally opposite to the velocity of the oscillating object and given by:

$$\vec{F} = -c\vec{v}$$

> Where F=the damping force
> c=the damping coefficient
> v= the object's velocity

In the real world, friction is a very common source of damping, but anything that speeds the release of energy from the oscillating system will cause damping. Some of the most familiar examples come from musical instruments. For instance, if a string is plucked or a cymbal struck, it will oscillate and produce sound. Over time, air resistance reduces the amplitude of the oscillation and we hear the sound decrease in volume. However, this process may take a fairly long time. If, instead, we want the sound to stop quickly, we may place a hand against the string or cymbal, thereby quickly stopping the vibration and returning the system to equilibrium. In this example, the hand serves as a damper, absorbing energy from the oscillating instrument.

Engineers and designers often design systems with appropriate damping mechanisms to minimize vibration and oscillations. The shock absorbers on cars are an example of such a damper. After driving over a bump, the car may "bounce up and down" (oscillate) for a long period of time, but the shock absorbers are designed to dissipate the energy quickly and shortened the time during which the car oscillates. If a damper allows an object, such as the car in this case, to return to equilibrium in the shortest possible time, the system is termed *critically damped*.

SUBAREA III.   ELECTRICITY AND MAGNETISM

COMPETENCY 11.0   UNDERSTAND ELECTRIC CHARGE, ELECTRIC FIELDS, AND ELECTRIC POTENTIAL

Skill 11.1   Describing the nature of charge

At its most fundamental level, electric charge is a conserved property of subatomic particles. Charge at this level is quantized as a multiple of the elementary charge e. Thus protons have a charge of +1 and electrons of -1. Electric charge is also manifest in macroscopic objects. In these objects, electric charge is the sum of the electric charges of its constituent particles. The SI unit for electric charge is the coulomb, defined as the quantity of charge that passes through the cross-section of a conductor carrying one ampere in one second.

In individual neutral atoms and the majority of macroscopic objects, the number of protons and electrons is balanced and thus a net neutral charge is observed. However, the charges with an object can be separated spatially, creating a polarization of charge. When charge flows through a material in a particular direction, it is known as an electric current. One of the most important properties of electric charges is that same-sign charged particles repel one another, while different-sign charged particles attract. This is known as Coulomb's law and is best understood by considering the behavior of point charges (see **Skill 11.3**). Additionally, it is important to understand that electrically charged materials are affected by and produce electromagnetic fields as described in **Competencies 13 and 14.**

Charges can be sensed using an electroscope, which can be a helpful heuristic tool for understanding some basic concepts surrounding positive and negative charges. A simple design for an electroscope involves a glass container with a metal ball outside and a metal contact inside. Connected to the contact is a thin metal leaf that can be deflected by only a small applied force.

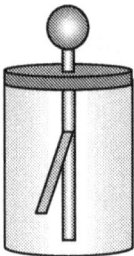

The behavior of the metal leaf inside the electroscope provides insight into the presence of charges in the surrounding environment or in the metal portions of the electroscope itself.

If a charged object is brought near (but does not touch) the electroscope (i.e., the metal ball), the free charge carriers in the metal rearrange themselves to minimize the energy of the system. If the object is negatively charged due to an excess of electrons, for example, the otherwise evenly distributed electrons in the metal of the electroscope will reorient such that a net positive charge is maintained near the charged object. This reorientation leads to a net negative charge in the other portion of the electroscope (i.e., inside the glass container), since the electroscope is initially uncharged. The local net charge results in electrical repulsion between the metal contact and the metal leaf. The leaf is then deflected.

If, while the charged object is near, the metal ball is touched by an observer, for example, the electroscope becomes "grounded," and some of the excess charge in certain areas is discharged. This results in the leaf returning to its original undeflected position. When the charged object is removed (after the observer also breaks contact), the result is that the metal leaf remains deflected. This phenomenon results from the excess charge obtained from grounding the electroscope during the time it was in the presence of a charged object. In the case of the above example, an excess of positive charge would remain and would disperse evenly throughout the metal (i.e., evenly very near the surface).

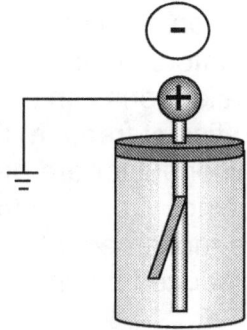 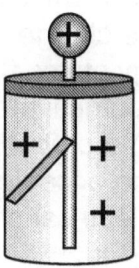

If the electroscope is grounded once more, with the charged object no longer present, the excess charge will be lost and the leaf will return to its undeflected position.

In addition to the above method of charging the electroscope by induction, it may also be charged by rubbing the outside metal with a charged object, such as a glass rod. The conducting metal allows the excess charge on the rod to further distribute itself into a preferred lower-energy situation. This results in a deflection of the leaf that, as with the previous case, can be reversed by grounding the electroscope.

Skill 11.2   Describing static charges in conductors and insulators

Electrical current requires the free flow of electrons. Various materials allow different degrees of electron movement and are classified as conductors, insulators, or semiconductors (in certain, typically man-made environments, superconductors also exist). When charge is transferred to a mass of material, the response is highly dependent on whether that material is a conductor or insulator.

Conductors: Those materials that allow for free and easy movement of electrons are called conductors. Some of the best conductors are metal, especially copper and silver. This is because these materials are held together with metallic bonds, which involve de-localized electrons shared by atoms in a lattice. If a charge is transferred to a conductor, the electrons will flow freely and the charge will quickly distribute itself across the material in a manner dictated by the conductor's shape.

Insulators: Materials that do not allow conduction are call insulators. Good insulators include, glass, rubber, and wood. These materials have chemical structures in which the electrons are closely localized to the individual atoms. In contrast to a conductor, a charge transferred to an insulator will remain localized at the point where it was introduced. This is because the movement of electrons will be highly impeded.

Semiconductors: Materials with intermediate conduction properties are known as semiconductors. Their properties are similar to insulators in that they have few free electrons to carry the charge. However, these electrons can be thermally excited into higher energy states that allow them sufficient freedom to transmit electrical charge. The electrical properties of a semiconductor are often improved by introducing impurities, a procedure known as doping. Doping introduces extra electrons or extra electron acceptors to facilitate the movement of charge. *Temporary* changes in electrical properties can be induced by applying an electrical field.

Charge can be transferred to materials in a few different ways:

Conduction: In the most general sense, electrical conduction is the movement of charged particles (electrons) through a medium. As explained above, in conducting materials, electrons loosely attached to atoms are capable of carrying an electrical current. This requires that the atoms of the conducting material be brought into physical contact with the charge source. For example, if a conductor is brought in contact with another charged conductor or a current source (such as a battery), the charge will be transferred to that conductor.

Friction: When two materials are brought into contact (or rubbed against each other), electrons may be transferred between them. If the materials are then separated, one will be left with a negative charge and the other will have a positive charge. This phenomenon is known as the triboelectric effect. Both the polarity and strength of the resultant charge depends upon the materials, surface roughness, temperature, and strain. Some common examples of combinations that produce significant charges are glass with silk and rubber with fur.

Induction: Electromagnetic induction is the production of voltage that occurs when a conductor interacts with a magnetic field. A conductor can be charged either by moving it through a static electric field or by placing it in a changing magnetic field. Induction was discovered by Michael Faraday and Faraday's Law governs this phenomenon.

Skill 11.3 Applying Coulomb's law to determine forces and fields due to various charge distributions

Any point charge may experience force resulting from attraction to or repulsion from another charged object. The easiest way to begin analyzing this phenomenon and calculating this force is by considering two point charges. Let us say that the charge on the first point is Q_1, the charge on the second point is Q_2, and the distance between them is r. Their interaction is governed by Coulomb's Law which gives the formula for the force F as:

$$F = k\frac{Q_1 Q_2}{r^2}$$

where $k = 9.0 \times 10^9 \frac{N \cdot m^2}{C^2}$ (known as Coulomb's constant)

The charge is a scalar quantity, however, the force has direction. For two point charges, the direction of the force is along a line joining the two charges. Note that the force will be repulsive if the two charges are both positive or both negative and attractive if one charge is positive and the other negative. Thus, a negative force indicates an attractive force.

When more than one point charge is exerting force on a point charge, we simply apply Coulomb's Law multiple times and then combine the forces as we would in any statics problem. Let's examine the process in the following example problem.

Problem: Three point charges are located at the vertices of a right triangle as shown below. Charges, angles, and distances are provided (drawing not to scale). Find the force exerted on the point charge A.

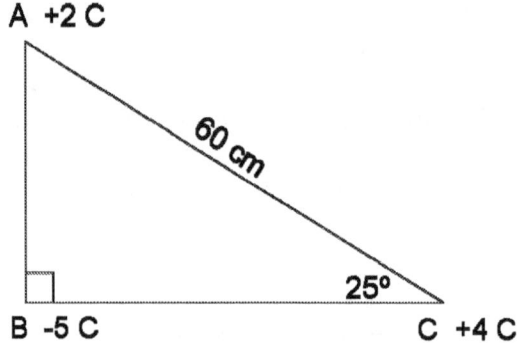

Solution: First we find the individual forces exerted on A by point B and point C. We have the information we need to find the magnitude of the force exerted on A by C.

$$F_{AC} = k\frac{Q_1 Q_2}{r^2} = 9 \times 10^9 \frac{N \cdot m^2}{C^2} \left(\frac{4C \times 2C}{(0.6m)^2} \right) = 2 \times 10^{11} N$$

To determine the magnitude of the force exerted on A by B, we must first determine the distance between them.

$$\sin 25° = \frac{r_{AB}}{60cm}$$
$$r_{AB} = 60cm \times \sin 25° = 25cm$$

Now we can determine the force.

$$F_{AB} = k\frac{Q_1 Q_2}{r^2} = 9 \times 10^9 \frac{N \cdot m^2}{C^2} \left(\frac{-5C \times 2C}{(0.25m)^2} \right) = -1.4 \times 10^{12} N$$

We can see that there is an attraction in the direction of B (negative force) and repulsion in the direction of C (positive force). To find the net force, we must consider the direction of these forces (along the line connecting any two point charges). We add them together using the law of cosines.

$$F_A^2 = F_{AB}^2 + F_{AC}^2 - 2F_{AB}F_{AC}\cos 75°$$
$$F_A^2 = (-1.4 \times 10^{12} N)^2 + (2 \times 10^{11} N)^2 - 2(-1.4 \times 10^{12} N)(2 \times 10^{11} N)^2 \cos 75°$$
$$F_A = 1.5 \times 10^{12} N$$

This gives us the magnitude of the net force, now we will find its direction using the law of sines.

$$\frac{\sin\theta}{F_{AC}} = \frac{\sin 75°}{F_A}$$

$$\sin\theta = F_{AC}\frac{\sin 75°}{F_A} = 2\times 10^{11}\,N\frac{\sin 75°}{1.5\times 10^{12}\,N}$$

$$\theta = 7.3°$$

Thus, the net force on A is 7.3° west of south and has magnitude 1.5×10^{12} N. Looking back at our diagram, this makes sense, because A should be attracted to B (pulled straight south) but the repulsion away from C "pushes" this force in a westward direction.

Electric Fields

Electric fields can be generated by a single point charge or by a collection of charges in close proximity. The electric field generated from a point charge is given by:

$$E = \frac{kQ}{r^2}$$

where E = the electric field

$k = 9.0 \times 10^9 \frac{N \cdot m^2}{C^2}$ (Coulomb's constant)

Q = the point charge
r = distance from the charge

Electric fields are visualized with field lines, which demonstrate the strength and direction of an electric field. The electric field around a positive charge points away from the charge and the electric field around a negative charge points toward the charge.

While it's easy enough to calculate and visualize the field generated by a single point charge, we can also determine the nature of an electric field produced by a collection of charge simply by adding the vectors from the individual charges. This is known as the superposition principle. The following equation demonstrates how this principle can be used to determine the field resulting from hundreds or thousands of charges.

$$\vec{E}_{total} = \sum_i \vec{E}_i = \vec{E}_1 + \vec{E}_2 + \vec{E}_3 \ldots$$

A field is essentially a parameter that has a certain value at each point in space. Fields may have only magnitude (scalar fields) or both magnitude and direction (vector fields).

Temperature is an example of a scalar field since it has a certain value that can be measured at any point. However, there is no directionality associated with temperature. A vector field like an electric field, however, is characterized both by a magnitude and a direction. The vector nature of the electric field arises from the fact that it is essentially the force that a unit positive charge would experience if placed in the field. Even though the field exists in the absence of any such actual charge, the magnitude and direction of the field are defined in this way.

Electric field lines, indicating the direction of the field at a point in space, have been discussed in the previous section. The previous section also discusses how multiple electric fields may be combined, like forces, using vector addition.

Skill 11.4 Applying the concepts of electrostatic potential energy, potential, and capacitance

Electric potential is simply the potential energy per unit of charge. Given this definition, it is clear that electric potential must be measured in joules per coulomb and this unit is known as a volt (J/C=V).

Within an electric field there are typically differences in potential energy. This potential difference may be referred to as voltage. The difference in electrical potential between two points is the amount of work needed to move a unit charge from the first point to the second point.

Stated mathematically, this is:

$$V = \frac{W}{Q}$$

where V= the potential difference
W= the work done to move the charge
Q= the charge

We know from mechanics, however, that work is simply force applied over a certain distance. We can combine this with Coulomb's law to find the work done between two charges distance r apart.

$$W = F \cdot r = k\frac{Q_1 Q_2}{r^2} \cdot r = k\frac{Q_1 Q_2}{r}$$

Now we can simply substitute this back into the equation above for electric potential:

$$V_2 = \frac{W}{Q_2} = \frac{k\frac{Q_1 Q_2}{r}}{Q_2} = k\frac{Q_1}{r}$$

Let's examine a sample problem involving electrical potential.

Problem: What is the electric potential at point A due to the 2 shown charges? If a charge of +2.0 C were infinitely far away, how much work would be required to bring it to point A?

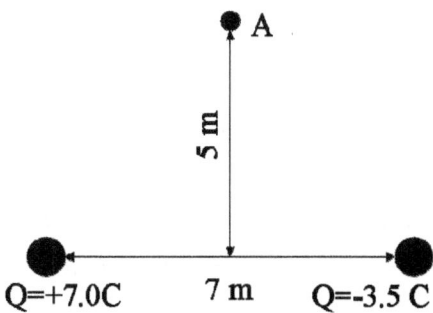

Solution: To determine the electric potential at point A, we simple find and add the potential from the two charges (this is the principle of superposition). From the diagram, we can assume that A is equidistant from each charge. Using the Pythagorean theorem, we determine this distance to be 6.1 m.

$$V = \frac{kq}{r} = k\left(\frac{7.0C}{6.1m} + \frac{-3.5C}{6.1m}\right) = 9 \times 10^9 \frac{N.m^2}{C^2}\left(0.57\frac{C}{m}\right) = 5.13 \times 10^9 V$$

Now, let's consider bringing the charged particle to point A. We assume that electric potential of these particle is initially zero because it is infinitely far away. Since now know the potential at point A, we can calculate the work necessary to bring the particle from V=0, i.e. the potential energy of the charge in the electrical field:

$$W = VQ = (5.13 \times 10^9) \times 2J = 10.26 \times 10^9 J$$

The large results for potential and work make it apparent how large the unit coulomb is. For this reason, most problems deal in microcoulombs (μC).

Capacitance
Capacitance (C) is a measure of the stored electric charge per unit electric potential. The mathematical definition is:

$$C = \frac{Q}{V}$$

It follows from the definition above that the units of capacitance are coulombs per volt, a unit known as a farad (F=C/V). In circuits, devices called parallel plate capacitors are formed by two closely spaced conductors. The function of capacitors is to store electrical energy.

When a voltage is applied, electrical charges build up in both the conductors (typically referred to as plates). These charges on the two plates have equal magnitude but opposite sign. The capacitance of a capacitor is a function of the distance d between the two plates and the area A of the plates:

$$C \approx \frac{\varepsilon A}{d}; A \gg d^2$$

Capacitance also depends on the permittivity of the non-conducting matter between the plates of the capacitor. This matter may be only air or almost any other non-conducting material and is referred to as a dielectric. The permittivity of empty space ε_0 is roughly equivalent to that for air, $\varepsilon_{air}=8.854 \times 10^{-12}$ C^2/N•m^2.

For other materials, the dielectric constant, κ, is the permittivity of the material in relation to air ($\kappa=\varepsilon/\varepsilon_{air}$). The make-up of the dielectric is critical to the capacitor's function because it determines the maximum energy that can be stored by the capacitor. This is because an overly strong electric field will eventually destroy the dielectric.

In summary, a capacitor is "charged" as electrical energy is delivered to it and opposite charges accumulate on the two plates. The two plates generate electric fields and a voltage develops across the dielectric. The energy stored in the capacitor, then, is equal to the amount of work necessary to create this voltage. The mathematical statement of this is:

$$E_{stored} = \frac{1}{2}CV^2 = \frac{1}{2}\frac{Q^2}{C} = \frac{1}{2}VQ$$

The work per unit volume or the electric field energy density within a capacitor can be shown to be $\eta = \frac{1}{2}\varepsilon E^2$. This result is generally valid for the energy per unit volume of any electrostatic field, not only for a constant field within a capacitor.

Problem: Imagine that a parallel plate capacitor has an area of 10.00 cm^2 and a capacitance of 4.50 pF. The capacitor is connected to a 12.0 V battery. The capacitor is completely charged and then the battery is removed. What is the separation of the plates in the capacitor? How much energy is stored between the plates? We've assumed that this capacitor initially had no dielectric (i.e., only air between the plates) but now imagine it has a Mylar dielectric that fully fills the space. What will the new capacitance be? (for Mylar, □=3.5)

Solution: To determine the separation of the plates, we use our equation for a capacitor:

$$C = \frac{\varepsilon_0 A}{d}$$

We can simply solve for d and plug in our values:

$$d = \varepsilon_0 \frac{A}{C} = \left(8.854 \times 10^{-12} \frac{C}{N \cdot m^2}\right) \frac{10 \times 10^{-4} m^2}{4.5 \times 10^{-12} F} = 1.97 \times 10^{-3} m = 1.97 mm$$

Similarly, to find stored energy, we simply employ the equation above:

$$E_{stored} = \frac{1}{2} QV$$

But we don't yet know the charge Q, so we must first find it from the definition of capacitance:

$$C = \frac{Q}{V}$$
$$Q = CV = (4.5 \times 10^{-12}) \times (12V) = 5.4 \times 10^{-11} C$$

Now we can find the stored energy:

$$E_{stored} = \frac{1}{2} QV = \frac{1}{2}(5.4 \times 10^{-11} C)(12V) = 3.24 \times 10^{-10} J$$

To find the capacitance with a Mylar dielectric, we again use the equation for capacitance of a parallel plate capacitor. Note that the new capacitance can be found by multiplying the original capacitance by κ:

$$C = \frac{\kappa_{Mylar} \varepsilon_0 A}{d} = \kappa_{Mylar} C_0 = 3.5 \times 4.5 pF = 15.75 pF$$

TEACHER CERTIFICATION STUDY GUIDE

COMPETENCY 12.0 UNDERSTAND SIMPLE CIRCUITS

Skill 12.1 Includes describing the properties of conductors, insulators, semiconductors, and Superconductors

Semiconductors, conductors and superconductors are differentiated by how "easily" current can flow in the presence of an applied electric field. Dielectrics (insulators) conduct little or no current in the presence of an applied electric field, as the charged components of the material (for example, atoms and their associated electrons) are tightly bound and are not free to move within the material. In the case of materials with some amount of conductivity, so-called valence (higher energy level) electrons are only loosely bound and may move among positive charge centers (i.e., atoms or ions).

Conductors, such as metals like aluminum and iron, have numerous valence electrons distributed among the atoms that compose the material. These electrons can move freely, especially under the influence of an applied electric field. In the absence of any applied field, these electrons move randomly, resulting in no net current. If a static electric field is applied, the mobile electrons reorient themselves to minimize the energy of the system and create an equipotential on the surface of the metal (in the ideal case of infinite conductivity).The highly mobile charge carriers do not allow a net field inside the metal; thus conductors can "shield" electromagnetic fields.

Semiconductors, such as silicon and germanium, are materials that can neither be described as conductors, nor as insulators. A certain number of charge carriers are mobile in the semiconductor, but this number is nowhere near the free charge populations of conductors such as metals. "Doping" of a semiconductor by adding so-called donor atoms or acceptor atoms to the intrinsic (or pure) semiconductor can increase the conductivity of the material. Furthermore, combination of a number of differently doped semiconductors (such as donor-doped silicon and acceptor-doped silicon) can produce a device with beneficial electrical characteristics (such as the diode). The conductivity of such devices can be controlled by applying voltages across specific portions of the device.

According to the theory of energy bands, as derived from quantum mechanics, electrons in the ground state reside in the valence band and are bound to their associated atoms or molecules. If the electrons gain sufficient energy, such as through heat, they can jump across the forbidden band (in which no electrons may exist) to the conduction band, thus becoming free electrons that can form a current in the presence of an applied field. For conductors, the valence band and conduction band meet or overlap, allowing electrons to easily jump to unoccupied conduction states. Thus, conductors have an abundance of free electrons. It is noteworthy that only two bands are shown in the diagram above, but that an infinite number of bands may exist at higher energies.

A more general statement of the difference in band structures is that, for insulators and semiconductors, the valence electrons fill up all the states in a particular band, leaving a gap between the highest energy valence electrons and the next available band. The difference between these two types of materials is simply a matter of the "size" of the forbidden band. For conductors, the band that contains the highest energy electrons has additional available states.

A useful way to visualize the relationship of conductors and semiconductors is by way of energy band diagrams. For purposes of comparison, an example of an insulator is included here as well.

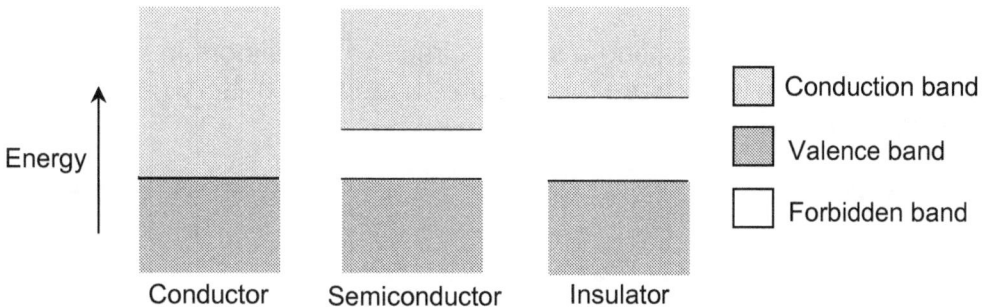

A nearly ideal conducting material is a superconductor. As a material increases in temperature, increased vibrational motion of the atoms or molecules leads to decreased charge carrier mobility and decreased conductivity. In the case of semiconductors, the increase in free carrier population outweighs the loss in mobility of the charge carriers, meaning that the semiconductor increases in conductivity as temperature increases. Superconductors, on the other hand, reach their peak conductivity at extremely low temperatures (although there are currently numerous efforts to achieve superconductivity at higher and higher temperatures, with room temperature or higher being the ultimate goal). The critical temperature of the material is the temperature at which superconducting properties emerge. At this temperature, the material has a nearly infinite conductivity and maintains an almost perfect equipotential across its surface when in the presence of a static electric field. Inside a superconductor, the electric field is virtually zero at all times. As a result, the time derivative of the electric flux density is zero, and, by Maxwell's equations, the magnetic flux density must likewise be zero.

$$\nabla \times \mathbf{H} = \frac{\partial \mathbf{D}}{\partial t} + \mathbf{J}$$

Since the electric field is also zero, the current density "J" inside the superconductor must also be zero (or very nearly so). This elimination of the magnetic flux density inside a superconductor is called the Meissner effect.

Skill 12.2 Applying Ohm's and Kirchhoff's laws to the analysis of series and parallel circuits

The two most important elements in simple circuits are resistors and capacitors. Often resistors and capacitors are used together in series or parallel. Two components are in series if one end of the first element is connected to one end of the second component. The components are in parallel if both ends of one element are connected to the corresponding ends of another. A series circuit has a single path for current flow through all of its elements. A parallel circuit is one that requires more than one path for current flow in order to reach all of the circuit elements.

Below is a diagram demonstrating a simple circuit with resistors in parallel (on right) and in series (on left). Note the symbols used for a battery (noted V) and the resistors (noted R).

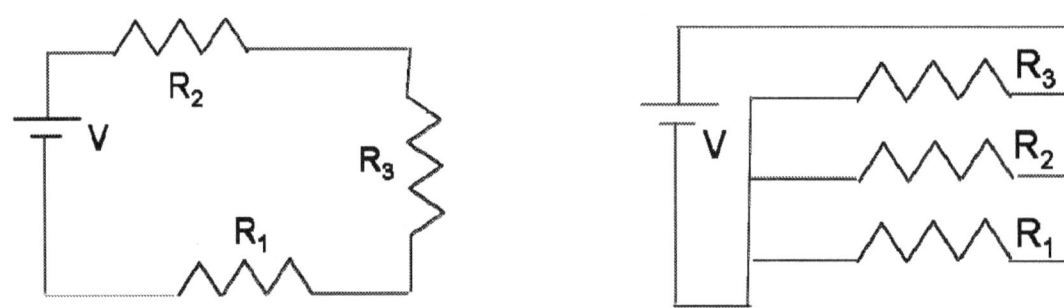

Thus, when the resistors are placed in series, the current through each one will be the same. When they are placed in parallel, the voltage through each one will be the same. To understand basic circuitry, it is important to master the rules by which the equivalent resistance (R_{eq}) or capacitance (C_{eq}) can be calculated from a number of resistors or capacitors:

Resistors in parallel: $\dfrac{1}{R_{eq}} = \dfrac{1}{R_1} + \dfrac{1}{R_2} + \cdots + \dfrac{1}{R_n}$

Resistors in series: $R_{eq} = R_1 + R_2 + \cdots + R_n$

Capacitors in parallel: $C_{eq} = C_1 + C_2 + \cdots + C_n$

Capacitors in series: $\dfrac{1}{C_{eq}} = \dfrac{1}{C_1} + \dfrac{1}{C_2} + \cdots + \dfrac{1}{C_n}$

TEACHER CERTIFICATION STUDY GUIDE

Kirchoff's Laws

Kirchoff's Laws are a pair of laws that apply to conservation of charge and energy in circuits and were developed by Gustav Kirchoff.

Kirchoff's Current Law: At any point in a circuit where charge density is constant, the sum of currents flowing toward the point must be equal to the sum of currents flowing away from that point.

Kirchoff's Voltage Law: The sum of the electrical potential differences around a circuit must be zero.

While these statements may seem rather simple, they can be very useful in analyzing DC circuits, those involving constant circuit voltages and currents.

Problem:
The circuit diagram at right shows three resistors connected to a battery in series. A current of 1.0 A is generated by the battery. The potential drop across R_1, R_2, and R_3 are 5V, 6V, and 10V. What is the total voltage supplied by the battery?

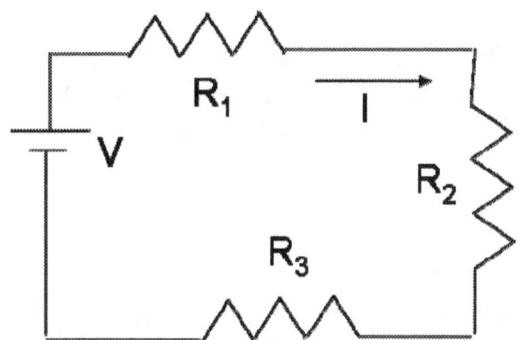

Solution:
Kirchoff's Voltage Law tells us that the total voltage supplied by the battery must be equal to the total voltage drop across the circuit. Therefore:

$$V_{battery} = V_{R_1} + V_{R_2} + V_{R_3} = 5V + 6V + 10V = 21V$$

Problem:
The circuit diagram at right shows three resistors wired in parallel with a 12V battery. The resistances of R_1, R_2, and R_3 are 4 Ω, 5 Ω, and 6 Ω, respectively. What is the total current?

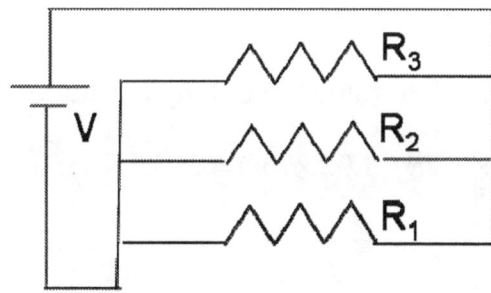

Solution:
This is a more complicated problem. Because the resistors are wired in parallel, we know that the voltage entering each resistor must be the same and equal to that supplied by the battery. We can combine this knowledge with Ohm's Law (see next section) to determine the current across each resistor:

$$I_1 = \frac{V_1}{R_1} = \frac{12V}{4\Omega} = 3A$$

$$I_2 = \frac{V_2}{R_2} = \frac{12V}{5\Omega} = 2.4A$$

$$I_3 = \frac{V_3}{R_3} = \frac{12V}{6\Omega} = 2A$$

Finally, we use Kirchoff's Current Law to find the total current:

$$I = I_1 + I_2 + I_3 = 3A + 2.4A + 2A = 7.4A$$

Ohm's Law

Ohm's Law states that the current passing through a conductor is directly proportional to the voltage drop and inversely proportional to the resistance of the conductor. Stated mathematically, this is:

$$V = IR$$

Problem:
The circuit diagram at right shows three resistors connected to a battery in series. A current of 1.0A flows through the circuit in the direction shown. It is known that the equivalent resistance of this circuit is 25 Ω. What is the total voltage supplied by the battery?

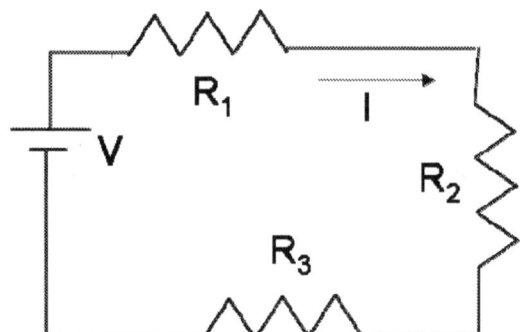

Solution:

To determine the battery's voltage, we simply apply Ohm's Law:

$$V = IR = 1.0A \times 25\Omega = 25V$$

Skill 12.3 Properly using voltmeters and ammeters

Electrical meters function by utilizing the following familiar equations:

Across a resistor (Resistor R): $V_R = IR_R$

Across a capacitor (Capacitor C): $V_C = IX_C$

Across an inductor (Inductor L): $V_L = IX_L$

Where V=voltage, I=current, R=resistance, X=reactance.

Ammeter: An ammeter placed in series in a circuit measures the current through the circuit. An ammeter typically has a very small resistance so that the current in the circuit is not changed too much by insertion of the ammeter.

Voltmeter: A voltmeter is used to measure potential difference. The potential difference across a resistor is measured by a voltmeter placed in parallel across it. An ideal voltmeter has very high resistance so that it does not appreciably alter circuit resistance and therefore the voltage drop it is measuring.

Galvanometer: A galvanometer is a device that measures current and is a component of an ammeter or a voltmeter. A typical galvanometer consists of a coil of wire, an indicator and a scale that is designed to be proportional to the current in the galvanometer. The principle that a current-carrying wire experiences a force in a magnetic field is used in the construction of a galvanometer. In order to create a voltmeter from a galvanometer, resistors are added in series to it. To build an ammeter using a galvanometer, a small resistance known as a **shunt resistor** is placed in parallel with it.

Potentiometer: A potentiometer is a variable resistance device in which the user can vary the resistance to control the current and voltage applied to a circuit. Since the potentiometer can be used to control what fraction of the emf of a battery is applied to a circuit, it is also known as a voltage divider. It can be used to measure an unknown voltage by comparing it with a known value.

Multimeter: A common electrical meter, typically known as a multimeter, is capable of measuring voltage, resistance, and current. Many of these devices can also measure capacitance (farads), frequency (hertz), duty cycle (a percentage), temperature (degrees), conductance (siemens), and inductance (henrys).

Skill 12.4 Determining power dissipated by circuit elements

Electrical power is a measure of how much energy is expended or how much work can be done by an electrical current and has units of watts (W). To determine electrical power, we simply use Joule's law:

$$P = IV$$

where P=power
I=current
V=voltage

If we combine this with Ohm's law (V=IR), we generate two new equations that are useful for finding the amount of power dissipated by a resistor:

$$P = I^2 R \qquad P = \frac{V^2}{R}$$

Problem: How much power is dissipated by a 1 kΩ resistor with a 50V voltage drop through it?

Solution: $P = \dfrac{V^2}{R} = \dfrac{(50V)^2}{1000\Omega} = 2.5W$

Skill 12.5 Analyzing energy transfer and conservation in electrical circuits

The first law of thermodynamics applies in a number of situations. Ideally, in many simple models of various phenomena, undesirable effects such as friction, resistance and absorption are ignored. In reality, however, these are unavoidable consequences of the nature of the phenomena. In the context of electrical circuits, collisions in wires or devices impedes the flow of electrons and can convert mechanical energy, which originates from an applied electric or magnetic field, into heat energy. If the heat is dissipated from the circuit then that heat loss represents energy loss, implying a decrease in the circuit's efficiency since less energy is available to do work.

COMPETENCY 13.0 UNDERSTAND MAGNETIC FIELDS

Skill 13.1 Describe the properties of permanent magnets

Magnetism is a phenomenon in which certain materials, known as magnetic materials, attract or repel each other. A magnet has two poles, a south pole and a north pole. Like poles repel while unlike poles attract. Magnetic poles always occur in pairs known as magnetic dipoles. One cannot isolate a single magnetic pole. If a magnet is broken in half, opposite poles appear at both sides of the break point so that one now has two magnets each with a south pole and a north pole. No matter how small the pieces a magnet is broken into, the smallest unit at the atomic level is still a dipole.

A large magnet can be thought of as one with many small dipoles that are aligned in such a way that apart from the pole areas, the internal south and north poles cancel each other out. Destroying this long range order within a magnet by heating or hammering can demagnetize it. The dipoles in a non-magnetic material are randomly aligned while they are perfectly aligned in a preferred direction in permanent magnets. In a ferromagnet, there are domains where the magnetic dipoles are aligned, however, the domains themselves are randomly oriented. A ferromagnet can be magnetized by placing it in an external magnetic field that exerts a force to line up the domains.

A magnet produces a magnetic field that exerts a force on any other magnet or current-carrying conductor placed in the field. Magnetic field lines are a good way to visualize a magnetic field. The distance between magnetic fields lines indicates the strength of the magnetic field such that the lines are closer together near the poles of the magnets where the magnetic field is the strongest. The lines spread out above and below the middle of the magnet, as the field is weakest at those points furthest from the two poles. The SI unit for magnetic field known as magnetic induction is Tesla(T) given by 1T = 1 N.s/(C.m) = 1 N/(A.m). Magnetic fields are often expressed in the smaller unit Gauss (G) (1 T = 10,000 G). Magnetic field lines always point from the north pole of a magnet to the south pole.

Magnetic field lines can be plotted with a magnetized needle that is free to turn in 3 dimensions. Usually a compass needle is used in demonstrations. The direction tangent to the magnetic field line is the direction the compass needle will point in a magnetic field. Iron filings spread on a flat surface or magnetic field viewing film which contains a slurry of iron filings are another way to see magnetic field lines.

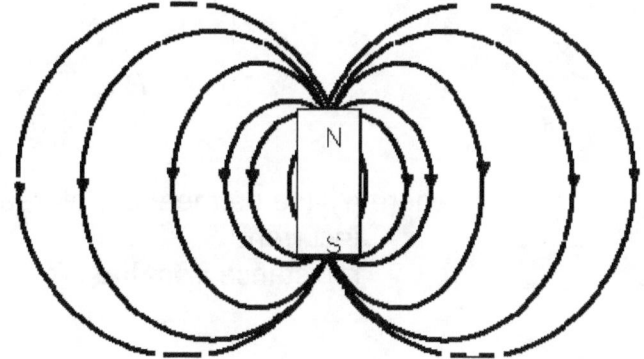

Skill 13.2 Applying laws to determine the orientation and strength of a magnetic field

Conductors through which electrical currents travel will produce magnetic fields: The magnetic field *dB* induced at a distance *r* by an element of current *Idl* flowing through a wire element of length *dl* is given by the **Biot-Savart** law

$$dB = \frac{\mu_0}{4\pi} \frac{Idl \times \hat{r}}{r^2}$$

where μ_0 is a constant known as the permeability of free space and $\hat{r}$ is the unit vector pointing from the current element to the point where the magnetic field is calculated.

An alternate statement of this law is **Ampere's law** according to which the line integral of *B.dl* around any closed path enclosing a steady current *I* is given by

$$\oint_C B \cdot dl = \mu_0 I$$

The basis of this phenomenon is the same no matter what the shape of the conductor, but we will consider three common situations:

Straight Wire
Around a current-carrying straight wire, the magnetic field lines form concentric circles around the wire. The direction of the magnetic field is given by the right-hand rule: When the thumb of the right hand points in the direction of the current, the fingers curl around the wire in the direction of the magnetic field. Note the direction of the current and magnetic field in the diagram.

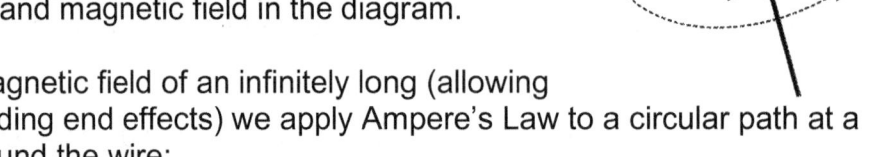

To find the magnetic field of an infinitely long (allowing us to disregarding end effects) we apply Ampere's Law to a circular path at a distance r around the wire:

$$B = \frac{\mu_0 I}{2\pi r}$$

where μ_0=the permeability of free space ($4\pi \times 10^{-7}$ T·m/A)
I=current
r=distance from the wire

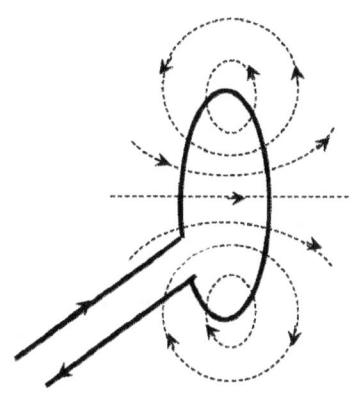

Loops

Like the straight wire from which it's been made, a looped wire has magnetic field lines that form concentric circles with direction following the right-hand rule. However, the field are additive in the center of the loop creating a field like the one shown. The magnetic field of a loop is found similarly to that for a straight wire.

In the center of the loop, the magnetic field is:

$$B = \frac{\mu_0 I}{2r}$$

Solenoids

A solenoid is essentially a coil of conduction wire wrapped around a central object. This means it is a series of loops and the magnetic field is similarly a sum of the fields that would form around several loops, as shown.

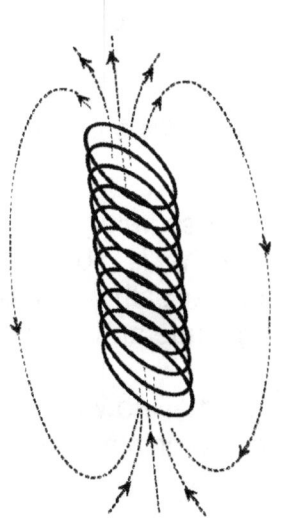

The magnetic field of a solenoid can be found as with the following equation:

$$B = \mu_0 n I$$

In this equation, n is turn density, which is simply the number of turns divided by the length of the solenoid.

TEACHER CERTIFICATION STUDY GUIDE

Skill 13.3 Determining the effect of a magnetic field on moving charges

The magnetic force exerted on a charge moving in a magnetic field depends on the size and velocity of the charge as well as the magnitude of the magnetic field. One important fact to remember is that only the velocity of the charge in a direction perpendicular to the magnetic field will affect the force exerted. Therefore, a charge moving parallel to the magnetic field will have no force acting upon it whereas a charge will feel the greatest force when moving perpendicular to the magnetic field.

The direction of the magnetic force is always at a right angle to the plane formed by the velocity vector v and the magnetic field B and is given by applying the right hand rule - if the fingers of the right hand are curled in a way that seems to rotate the v vector into the B vector, the thumb points in the direction of the force. The magnitude of the force is equal to the cross product of the velocity of the charge with the magnetic field multiplied by the magnitude of the charge.

$$F = q(v \times B) \quad \text{or} \quad F = qvB\sin(\theta)$$

Where θ is the angle formed between the vectors of velocity of the charge and direction of magnetic field.

Problem: Assuming we have a particle of 1×10^{-6} kg that has a charge of -8 coulombs that is moving perpendicular to a magnetic field in a clockwise direction on a circular path with a radius of 2 m and a speed of 2000 m/s, let's determine the magnitude and direction of the magnetic field acting upon it.

Solution: We know the mass, charge, speed, and path radius of the charged particle. Combining the equation above with the equation for centripetal force we get

$$qvB = \frac{mv^2}{r} \quad \text{or} \quad B = \frac{mv}{qr}$$

Thus B= $(1 \times 10^{-6}$ kg$)$ $(2000$ m/s$)$ / $(-8$ C$)(2$ m$)$ = 1.25×10^{-4} Tesla

Since the particle is moving in a clockwise direction, we use the right hand rule and point our fingers clockwise along a circular path in the plane of the paper while pointing the thumb towards the center in the direction of the centripetal force. This requires the fingers to curl in a way that indicates that the magnetic field is pointing out of the page. However, since the particle has a negative charge we must reverse the final direction of the magnetic field into the page.

A mass spectrometer measures the mass to charge ratio of ions using a setup similar to the one described above. m/q is determined by measuring the path radius of particles of known velocity moving in a known magnetic field.

PHYSICS

Skill 13.4 Explaining the role of magnetic force and torque in the operation of technological devices

Electromagnetism is the foundation for a vast number of modern technologies ranging from computers to communications equipment. More mundane technologies such as motors and generators are also based upon the principles of electromagnetism. The particular understanding of electrodynamics can either be in terms of quantum mechanics (quantum electrodynamics) or classical electrodynamics, depending on the type of phenomenon being analyzed. In classical electrodynamics, which is a sufficient approximation for most situations, the electric field, resulting from electric charge, and the magnetic field, resulting from moving charges, are the parameters of interest and are related through Maxwell's equations.

Motors
Electric motors are found in many common appliances such as fans and washing machines. The operation of a motor is based on the principle that a magnetic field exerts a force on a current carrying conductor. This force is essentially due to the fact that the current carrying conductor itself generates a magnetic field; the basic principle that governs the behavior of an electromagnet. In a motor, this idea is used to convert electrical energy into mechanical energy, most commonly rotational energy. Thus the components of the simplest motors must include a strong magnet and a current-carrying coil placed in the magnetic field in such a way that the force on it causes it to rotate.

A typical motor is composed of a stationary portion, called the stator, and a rotating (or moving) portion, called the rotor. Coils of wire that serve as electromagnets are wound on the armature, which can be either the stator or the rotor, and are powered by an electric source. Motors use electric current to generate a magnetic field around an electromagnet, which results in a rotational force due to the presence of an external magnetic field (either from permanent magnets or electromagnets). The designs of various motors can differ dramatically, but the general principles of electromagnetism that describe their operation are generally the same.

Generators
Generators are in effect "reverse motors". They exploit electromagnetic induction to generate electricity. Thus, if a coil of wire is rotated in a magnetic field, an alternating EMF is produced which allows current to flow. Any number of energy sources can be used to rotate the coil, including combustion, nuclear fission, flowing water or other sources. Therefore, generators are devices that convert mechanical or other forms of energy into electrical energy.

Meters

A number of different types of meters use electromagnetism or are designed to measure certain electromagnetic parameters. For example, older forms of ammeters (galvanometers), when supplied with a current, provided a measurement through the deflection of a spring-loaded needle. A coil connected to the needle acted as an electromagnet which, in the presence of a permanent magnetic field, would be deflected in the same manner as a rotor, as mentioned previously. The strength of the electromagnet, and thus the extent of the deflection, is proportional to the current. Further, the spring limits the deflection in such a manner that a reasonably accurate measurement of the current is provided.

Magnetic Media

Although the cassette tape has fallen out of favor with popular culture, magnetic media are still widely used for information storage. Magnetic strips on the back of credit and identification cards, computer hard drive disks and magnetic tapes (such as those contained in cassettes) are all examples of magnetic media. The principle underlying magnetic media is the sequential magnetization of a region of the medium. A special head is able to detect spatial magnetic fluctuations in the medium which are then converted into an electrical signal. The head is often able to "write" to the medium as well. The electrical signal from the medium can be converted into sound or video, as with the video or audio cassette player, or it can be digitized for use in a computer.

COMPETENCY 14.0 UNDERSTAND ELECTROMAGNETIC INDUCTION

Skill 14.1 Finding the rate of change of magnetic flux through a surface

When the magnetic flux through a coil is changed, a voltage is produced which is known as induced electromagnetic force. Magnetic flux is a term used to describe the number of magnetic fields lines that pass through an area and is described by the equation: $\Phi = BA\cos\theta$ (Where Φ is the angle between the magnetic field B, and the normal to the plane of the coil of area A)

Consider a coil lying flat on the page with a square cross section that is 10 cm by 5 cm. The coil consists of 10 loops and has a magnetic field of 0.5 T passing through it coming out of the page. Let's find the induced EMF when the magnetic field is changed to 0.8 T in 2 seconds.

First, let's find the initial magnetic flux: Φ_i

$$\Phi_i = BA\cos\theta = (.5\text{ T})(.05\text{ m})(.1\text{ m})\cos 0° = 0.0025\text{ T m}^2$$

And the final magnetic flux: Φ_f

$$\Phi_f = BA\cos\theta = (0.8\text{ T})(.05\text{ m})(.1\text{ m})\cos 0° = 0.004\text{ T m}^2$$

The induced emf is calculated then by

$$\varepsilon = -N\Delta\Phi/\Delta t = -10(.004\text{ T m}^2 - .0025\text{ T m}^2)/2\text{ s} = -0.0075\text{ volts}.$$

Skill 14.2 Analyzing factors that affect the magnitude of an induced emf

By changing any of the three inputs, magnetic field, area of coil, or angle between field and coil, the flux will change and an EMF can be induced. The speed at which these changes occur also affects the magnitude of the EMF, as a more rapid transition generates more EMF than a gradual one. This is described by **Faraday's law** of induction: $\varepsilon = -N\Delta\Phi/\Delta t$ (where ε is emf induced, N is the number of loops in a coil, t is time, and Φ is magnetic flux)

The negative sign signifies **Lenz's law** which states that induced emf in a coil acts to oppose any change in magnetic flux. Thus the current flows in a way that creates a magnetic field in the direction opposing the change in flux.

Skill 14.3 Determining the direction of an induced current or emf

The negative sign in Faraday's Law (described in the previous section) leads to Lenz's law which states that the induced current must produce a magnetic field that opposes the change in the applied magnetic field. This is simply an expression of the conservation of energy.

The right- or left-hand rule applies only in the case of a particular current convention: the right-hand rule is typically applied in the case of the positive current convention (i.e., the direction of flow of positive charge). Since positive and negative current are complementary, however, the rules for negative current simply involve using the left hand instead of the right hand. For simplicity, only the right-hand rule will be discussed explicitly.

The right-hand rule states that if the fingers of the right hand are curled in the direction of the positive current flow, the resulting magnetic flux is in the direction of the extended thumb. According to Lenz's law, when the magnetic flux through the surface bounded by the conductor increases, the induced current, $i_{induced}$, must produce a magnetic flux that is in the opposite direction to that of the applied flux. This is illustrated below.

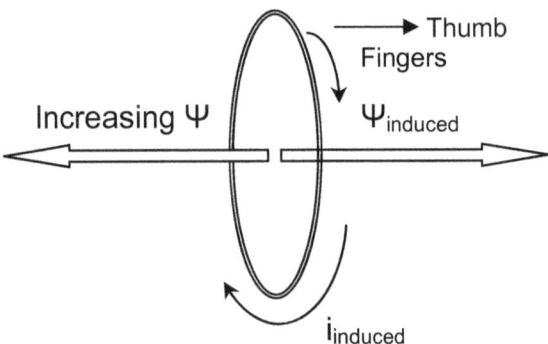

When the applied magnetic flux through the surface decreases, the direction of the induced flux must likewise coincide with the direction of the applied flux.

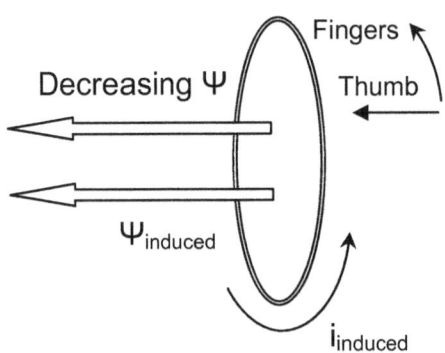

Thus, the induced magnetic flux must oppose the change in the applied flux. The direction of the induced positive current is the same as the direction of the curled fingers of the right hand when the thumb is extended in the direction of the induced magnetic flux. If negative current is of interest (for example, the flow of electrons), then the left-hand must be substituted in place of the right hand; the thumb follows the same rules, but the direction of the curled fingers indicates the direction of negative current flow.

Analyzing Lenz's Law in Terms of Conservation of Energy

Though it may not be entirely obvious initially, Lenz's law is a consequence of the conservation of energy. We can understand this by considering the physical scenario in which a permanent magnet (grey in diagram) moves through a loop of wire in the direction shown. An electric current will be induced in the wire and its direction will depend on which pole of the magnet enters the loop first. According to Lenz's law, the current will flow in the direction shown in this diagram. The direction of the current shown would be reversed if the magnet were turned so that the south pole entered first or if the magnet traveled in the direction opposite that shown by the arrow.

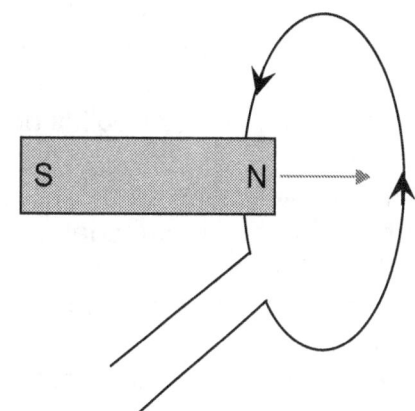

Now, suppose that Lenz's law did not hold and instead the magnet in the situation above induced a current in the direction opposite from that shown. That induced current would then induce a magnetic field directed toward the left of the page. This would, in turn, produce an attractive force between the permanent magnet and the induced magnetic field. A feedback loop would emerge: the permanent magnet would be drawn into the loop more quickly, a greater current induced, a stronger field generated, and the magnet drawn even more quickly into the loop. This would have an effect of increasing the kinetic energy of the magnet while also increasing the rate of energy dissipation of the loop. In short, energy would be created. Since this violation of the law of energy conservation cannot be correct, Lenz's law is supported.

Skill 14.4 Recognizing that magnetic energy is stored in an inductor

Inductance L measures the amount of magnetic flux Φ created by a current i and is given by $L = \dfrac{\Phi}{i}$. The unit of inductance is Henrys when the flux is expressed in Webers and the current in Amperes. While a capacitor resists a change in voltage through a circuit, an inductor resists a change in current. As indicated by Faraday's law, the magnetic flux generated by the inductor creates an emf that opposes the flow of current.

In a DC (direct current) circuit, a capacitor behaves like an open circuit (infinite resistance) while an inductor behaves like a short circuit (zero resistance) in the steady state following the initial transient response (where a capacitor builds up charge and an inductor builds up a magnetic field).

In an AC (alternating current) circuit, however, both capacitors and inductors contribute to the net impedance in the circuit which is a measure of opposition to the alternating current. It is similar to resistance and also has the unit ohm. However, due to the phased nature of AC, impedance is a complex number, having both real and imaginary components. The resistance is the real part of impedance while the reactance of capacitors and inductors constitute the imaginary part. The relationship between impedance (Z), resistance(R), and reactance (X) is given by below.

Skill 14.5 Describing alternators and the basic properties of alternating current

Alternating current (AC) is a type of electrical current with cyclically varying magnitude and direction. This is differentiated from direct current (DC), which has constant direction. AC is the type of current delivered to businesses and residences.

Though other waveforms are sometimes used, the vast majority of AC current is sinusoidal. Thus we can use wave terminology to help us describe AC current. Since AC current is a function of time, we can express it mathematically as:

$$v(t) = V_{peak} \cdot \sin(\omega t)$$

where V_{peak}= the peak voltage; the maximum value of the voltage
ω=angular frequency; a measure of rotation rate
t=time

TEACHER CERTIFICATION STUDY GUIDE

A few more terms are useful to help us characterize AC current:

Peak-to-peak value: The difference between the positive and negative peak values. Thus peak-to-peak value is equal to 2 x V_{peak}.

Root mean square value (V_{rms}, I_{rms}): A specific type of average found by the following formulae:

$$V_{rms} = \frac{V_{peak}}{\sqrt{2}} \;;\; I_{rms} = \frac{I_{peak}}{\sqrt{2}} \;;\; I_{rms} = \frac{V_{rms}}{R}$$

V_{rms} is useful because an AC current will deliver the same power as a DC current if its $V_{rms}=V_{DC}$, i.e. average power $P_{av} = V_{rms} I_{rms}$.

Frequency: Describes how often the wave passes through a particular point per unit time. Note that this is physical frequency, f, which is related to the angular frequency ω by:

$$\omega = 2\pi f$$

Impedance: A measure of opposition to an alternating current. It is similar to resistance and also has the unit ohm. However, due to the phased nature of AC, impedance is a complex number, having both real and imaginary components. The resistance is the real part of impedance while the reactance of capacitors and inductors constitute the imaginary part.

Resonant frequency: The frequency at which the impedance between the input and output of the circuit is minimum. At this frequency a phenomenon known as electrical resonance occurs.

Reactance: The impedance contributed by inductors and capacitors in AC circuit. Mathematically, reactance is the imaginary part of impedance. The relationship between impedance (Z), resistance(R), and reactance (X) is given by below.

$$Z = R + Xi$$

1. Remember that $i=\sqrt{-1}$

Problem:

An AC current has V_{rms}=220 V. What is its peak-to-peak value?

PHYSICS 99

Solution:

We simply determine V$_{peak}$ from the definition of V$_{rms}$:

$$V_{rms} = \frac{V_{peak}}{\sqrt{2}}$$

$$V_{peak} = V_{rms} \times \sqrt{2} = 220V \times \sqrt{2} = 311.12V$$

Therefore,

$$V_{peak-to-peak} = 2 \times V_{peak} = 2 \times 311.12V = 622.24V$$

Skill 14.6 Using the principle of electromagnetic induction to explain the operation of technological devices

Electromagnetic induction is used in a transformer, a device that magnetically couples two circuits together to allow the transfer of energy between the two circuits without requiring motion. Typically, a transformer consists of a couple of coils and a magnetic core. A changing voltage applied to one coil (the primary) creates a flux in the magnetic core, which induces voltage in the other coil (the secondary). All transformers operate on this simple principle though they range in size and function from those in tiny microphones to those that connect the components of the US power grid.

One of the most important functions of transformers is that they allow us to "step-up" and "step-down" between vastly different voltages. To determine how the voltage is changed by a transformer, we employ any of the following relationships:

$$\frac{V_s}{V_p} = \frac{n_s}{n_p} = \frac{I_p}{I_s}$$

where V$_s$=secondary voltage
V$_p$=primary voltage
n$_s$=number of turns on secondary coil
n$_p$=number of turns on primary coil
I$_p$=primary current
I$_s$=secondary current

Problem: If a step-up transformer has 500 turns on its primary coil and 800 turns on its secondary coil, what will be the output (secondary) voltage be if the primary coil is supplied with 120 V?

Solution:

$$\frac{V_s}{V_p} = \frac{n_s}{n_p}$$

$$V_s = \frac{n_s}{n_p} \times V_p = \frac{800}{500} \times 120V = 192V$$

Motors
Electric motors are found in many common appliances such as fans and washing machines. The operation of a motor is based on the principle that a magnetic field exerts a force on a current carrying conductor. This force is essentially due to the fact that the current carrying conductor itself generates a magnetic field; the basic principle that governs the behavior of an electromagnet. In a motor, this idea is used to convert electrical energy into mechanical energy, most commonly rotational energy. Thus the components of the simplest motors must include a strong magnet and a current-carrying coil placed in the magnetic field in such a way that the force on it causes it to rotate.

Motors may be run using DC or AC current and may be designed in a number of ways with varying levels of complexity. A very basic DC motor consists of the following components:

- A field magnet
- An armature with a coil around it that rotates between the poles of the field magnet
- A power supply that supplies current to the armature
- An axle that transfers the rotational energy of the armature to the working parts of the motor
- A set of commutators and brushes that reverse the direction of power flow every half rotation so that the armature continues to rotate

Generators
Generators are devices that are the opposite of motors in that they convert mechanical energy into electrical energy. The mechanical energy can come from a variety of sources; combustion engines, blowing wind, falling water, or even a hand crank or bicycle wheel. Most generators rely upon electromagnetic induction to create an electrical current. These generators basically consist of magnets and a coil. The magnets create a magnetic field and the coil is located within this field. Mechanical energy, from whatever source, is used to spin the coil within this field. As stated by Faraday's Law, this produces a voltage.

TEACHER CERTIFICATION STUDY GUIDE

SUBAREA IV.   WAVES, ACOUSTICS, AND OPTICS

COMPETENCY 15.0   UNDERSTAND THE CHARACTERISTICS OF WAVES AND WAVE MOTION

Skill 15.1   Includes describing the transfer of momentum and energy by wave motion

Waves are simply disturbances that propagate through space and time. Most waves must propagate through a medium, which may be solid, liquid, or gas. However, some electromagnetic waves do not require a medium and can move through a vacuum. Most waves transfer energy from their source to their destination, typically without actually transferring molecules of the medium through which they travel. The individual particles of the medium are temporarily displaced but eventually return to their equilibrium positions. As the particles oscillate, they transfer energy and momentum to a neighboring particle which, in turn, moves back and forth around its equilibrium position while transferring energy and momentum to another particle and so on.

All around us, there are examples of waves including ocean waves, sound, radio waves, microwaves, seismic waves, sunlight, x-rays, and radioactive gamma rays. Whether these waves actually displace media or simply carry energy, their positions fluctuate as they move through time and space. Often these fluctuations are regular and we can use both diagrams and mathematical equations to understand the pattern a wave follows in space and how quickly it moves.

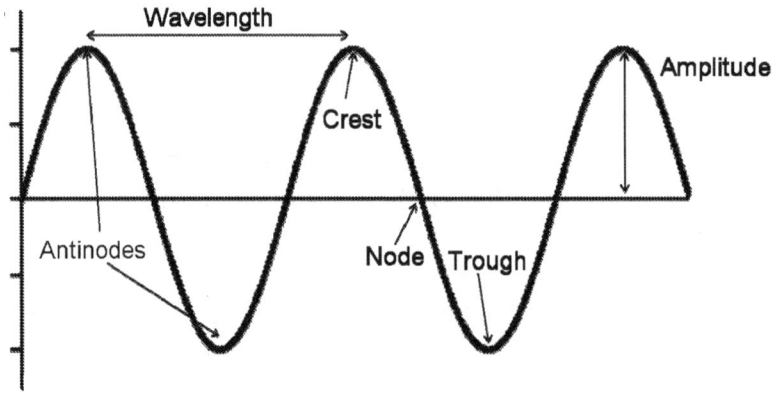

This diagram of a sinusoidal wave shows us displacement caused by the wave as it propagates through a medium. This displacement can be graphed against either time or distance. Note how displacement depends on the distance which the wave has traveled/ how much time has elapsed. So if we chose a particular displacement (let's say the crest), the wave will return to that displacement value (i.e., crest again) after one period (T) or one wavelength (λ).

PHYSICS   102

Skill 15.2 Comparing longitudinal and transverse waves

To fully understand waves, it is important to understand many of the terms used to characterize them.

Wave velocity: Two velocities are used to describe waves. The first is phase velocity, which is the rate at which a wave propagates. For instance, if you followed a single crest of a wave, it would appear to move at the phase velocity. The second type of velocity is known as group velocity and is the speed at which variations in the wave's amplitude shape propagate through space. Group velocity is often conceptualized as the velocity at which energy is transmitted by a wave. Phase velocity is denoted v_p and group velocity is denoted v_g.

Crest: The maximum value that a wave assumes; the highest point.

Trough: The lowest value that a wave assumes; the lowest point.

Nodes: The points on a wave with minimal amplitude.

Antinodes: The farthest point from the node on the amplitude axis; both the crests and the troughs are antinodes.

Amplitude: The distance from the wave's highest point (the crest) to the equilibrium point. This is a measure of the maximum disturbance caused by the wave and is typically denoted by A.

Wavelength: The distance between any two sequential troughs or crests denoted λ and representing a complete cycle in the repeated wave pattern.

Period: The time required for a complete wavelength or cycle to pass a given point. The period of a wave is usually denoted T.

Frequency: The number of periods or cycles per unit time (usually a second). The frequency is denoted f and is the inverse of the wave's period (that is, $f=1/T$).

Phase: This is a given position in the cycle of the wave. It is most commonly used in discussing a "being out of phase" or a "phase shift", an offset between waves.

Longitudinal waves: These are created by oscillations in the direction in which the wave travels. Thus, if we imagine a longitudinal waveform moving down a tube, particles will move back and forth parallel to the sides of the tube.

Transverse waves: These waves, on the other hand, oscillate in a direction perpendicular to the direction of wave travel. So let's imagine the same tube, this time with a transverse wave traveling down it. In this case, the particles oscillate up and down or side to side within the tube. Particle displacement in a transverse wave can also be easily visualized in the vibration of a taut string.

Polarization: A property of transverse waves that describes the plane perpendicular to the direction of travel in which the oscillation occurs. Note that longitudinal waves are not polarized because they can oscillate only in one direction, the direction of travel.

Skill 15.3   Analyzing and relating the characteristics of waves

The phase velocity of a wave is related to its wavelength and frequency. Taking light waves, for instance, the speed of light c is equal to the distance traveled divided by time taken. Since the light wave travels the distance of one wavelength λ in the period of the wave T,

$$c = \frac{\lambda}{T}$$

The frequency of a wave, f, is the number of completed periods in one second. In general,

$$f = \frac{1}{T}$$

So the formula for the speed of light can be rewritten as

$$c = \lambda f$$

Thus the velocity of a wave is equal to the wavelength times the frequency.

Skill 15.4   Explaining reflection, refraction, diffraction, and the Doppler effect

Wave refraction is a change in direction of a wave due to a change in its speed. This most commonly occurs when a wave passes from one material to another, such as a light ray passing from air into water or glass. However, light is only one example of refraction; any type of wave can undergo refraction. Another example would be physical waves passing from water into oil. At the boundary of the two media, the wave velocity is altered, the direction changes, and the wavelength increases or decreases. However, the frequency remains constant.

The index of refraction, n, is the amount by which light slows in a given material and is defined by the formula

$$n = \frac{c}{v}$$

where v represents the speed of light through the given material.

Problem: The speed of light in an unknown medium is measured to be $1.24 \times 10^8 \, m/s$. What is the index of refraction of the medium?

Solution:

$$n = \frac{c}{v}$$

$$n = \frac{3.00 \times 10^8}{1.24 \times 10^8} = 2.42$$

Referring to a standard table showing indices of refraction, we would see that this index corresponds to the index of refraction for diamond.

Reflection is the change in direction of a wave at an interface between two dissimilar media such that the wave returns into the medium from which it originated. The most common example of this is light waves reflecting from a mirror, but sound and water waves can also be reflected. The law of reflection states that the angle of incidence is equal to the angle of reflection.

Dispersion is the separation of a wave into its constituent wavelengths due to interaction with a material occurring in a wavelength-dependent manner (as in thin-film interference for instance). Dispersive prisms separate white light into these constituent colors by relying on the differences in refractive index that result from the varying frequencies of the light. Prisms rely on the fact that light changes speed as it moves from one medium to another. This then causes the light to be bent and/or reflected. The degree to which bending/reflection occurs is a function of the light's angle of incidence and the refractive indices of the media.

The Doppler effect is observed with all types of electromagnetic radiation as in the case of sound. The frequency or wavelength shift in the case of electromagnetic radiation is observed as a change in color. The equations, however, are only identical to that of sound when the relative speed of source and observer is much less than that of light.

The Doppler effect for light has been exploited by astronomers to measure the speed at which stars and galaxies are approaching or moving away. In astronomy, Doppler effect is measured in terms of wavelength rather than frequency. An approximate equation in cases where the relative velocity of source and observer is very much smaller than light speed c is $U = \Delta\lambda / \lambda$

"U" is the relative velocity of the source and observer, $\Delta\lambda$ is the change in wavelength of the observed light and λ is the actual wavelength of the light. If the wavelength decreases this is known as a **blue shift** since the frequency increases and the highest visible frequencies are in the blue range. An increase in wavelength is known as a **red shift.**

Skill 15.5 Applying the principle of superposition to investigate the properties of constructive and destructive interference

According to the principle of linear superposition, when two or more waves exist in the same place, the resultant wave is the sum of all the waves, i.e. the amplitude of the resulting wave at a point in space is the sum of the amplitudes of each of the component waves at that point. Interference is usually observed in coherent waves, well-correlated waves that have very similar frequencies or even come from the same source.

Superposition of waves may result in either constructive or destructive interference. Constructive interference occurs when the crests of the two waves meet at the same point in time. Conversely, destructive interference occurs when the crest of one wave and the trough of the other meet at the same point in time. It follows, then, that constructive interference increases amplitude and destructive interference decreased it. We can also consider interference in terms of wave phase; waves that are out of phase with one another will interfere destructively while waves that are in phase with one another will interfere constructively. In the case of two simple sine waves with identical amplitudes, for instance, amplitude will double if the waves are exactly in phase and drop to zero if the waves are exactly 180° out of phase.

Additionally, interference can create a standing wave, a wave in which certain points always have amplitude of zero. Thus, the wave remains in a constant position. Standing waves typically results when two waves of the same frequency traveling in opposite directions through a single medium are superposed. View an animation of how interference can create a standing wave at the following URL:

http://www.glenbrook.k12.il.us/GBSSCI/PHYS/mmedia/waves/swf.html

All wavelengths in the EM spectrum can experience interference but it is easy to comprehend instances of interference in the spectrum of visible light. One classic example of this is Thomas Young's double-slit experiment. In this experiment a beam of light is shone through a paper with two slits and a striated wave pattern results on the screen. The light and dark bands correspond to the areas in which the light from the two slits has constructively (bright band) and destructively (dark band) interfered.

Similarly, we may be familiar with examples of interference in sound waves. When two sounds waves with slightly different frequencies interfere with each other, beat results. We hear a beat as a periodic variation in volume with a rate that depends on the difference between the two frequencies. You may have observed this phenomenon when listening to two instruments being tuned to match; beating will be heard as the two instruments approach the same note and disappear when they are perfectly in tune.

COMPETENCY 16.0 UNDERSTAND THE PRINCIPLES OF SOUND AND ACOUSTICS

Skill 16.1 Includes explaining the production and propagation of sound waves

Sound waves are mechanical waves that can travel through various kinds of media, solid, liquid and gas. They are longitudinal waves transmitted as variations in the pressure of the surrounding medium. These variations in pressure form waves that are termed sound waves or acoustic waves. When the particles of the medium are drawn close together it is called compression. When the particles of the medium are spread apart it is known as rarefaction.

Sound waves have different characteristics in different materials. Boundaries between different media can result in partial or total reflection of sound waves. Thus, the phenomenon of echo takes place when a sound wave strikes a material of differing characteristics than the surrounding medium. The reflection of a voice off the walls of a room is a particular example.

The frequency of the harmonic acoustic wave, ω, determines the pitch of the sound in the same manner that the frequency determines the color of an electromagnetic harmonic wave. A high pitch sound corresponds to a high frequency sound wave and a low pitch sound corresponds to a low frequency sound wave. Similarly, the amplitude of the pressure determines the loudness or just as the amplitude of the electric (or magnetic) field E determines the brightness or intensity of a color. The intensity of a sound wave is proportional to the square of its amplitude.

Skill 16.2 Applying the principles of standing waves to explain resonance and to analyze the production of musical sounds in vibrating strings and air columns

A standing wave typically results from the interference between two waves of the same frequency traveling in opposite directions. The result is a stationary vibration pattern. One of the key characteristics of standing waves is that there are points in the medium where no movement occurs. The points are called nodes and the points where motion is maximal are called antinodes. Thus the amplitude of a standing wave varies with location. This property allows for the analysis of various typical standing waves.

Strings

Imagine a string of length L tied tightly at its two ends. We can generate a standing wave by plucking the string. The waves traveling along the string are reflected at the fixed end points and interfere with each other to produce standing waves. There will always be two nodes at the ends where the string is tied. Depending on the frequency of the wave that is generated, there may also be other nodes along the length of the string. In the diagrams below, examples are given of strings with 0, 1, or 2 additional nodes.

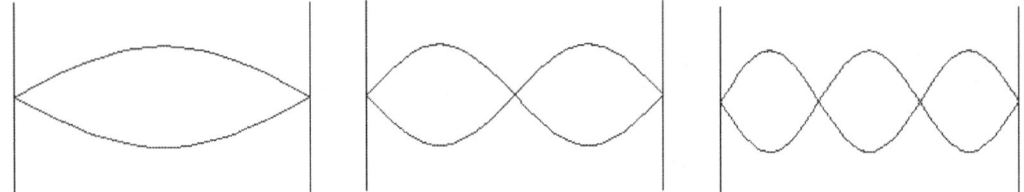

Percussion

On a percussion instrument such as a drum, standing waves occur on the two-dimensional surface in many different configurations. The nodes are circles and straight lines rather than points as on a string. Some possible node patterns are shown below.

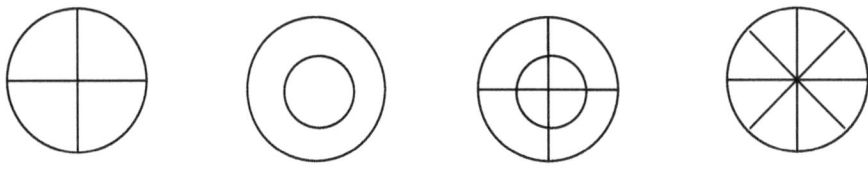

Winds

Just as on a string, standing waves can propagate in gaseous or liquid medium inside a tube. In these cases we will also observe harmonic vibrations, but their nature will depend on whether the ends of the tube are closed or open. Specifically, an antinode will be observed at an open end and a node will appear at a closed end. As in strings, there may be additional nodes in between.
Due to the presence of nodes in a standing wave pattern, a stretched string or a tube with ends closed vibrates at certain characteristic frequencies that correspond to different node patterns. These frequencies are known as **harmonics** or characteristic **resonance frequencies** of the instrument. The instrument will absorb much greater energy when excited at one of the resonance frequencies. The lowest of these frequencies is known as the **fundamental** while the higher ones are known as **overtones**. Harmonic frequencies for vibrating strings and air tubes are calculated below.

Vibrating string

Since we know the length of the string (L) in each case, we can calculate the wavelength (λ) and frequency (f) for any harmonic using the following formula, where n=the harmonic order (n=1,2,3...) and *v* is the phase velocity of the wave.

$$\lambda_n = \frac{2L}{n} \qquad f_n = \frac{v}{\lambda_n} = n\frac{v}{2L}$$

Waves in a tube

For a tube with two closed ends or two open ends (note that this is the same as for the string described above):

$$f_n = \frac{nv}{2L}$$

where n = 1,2,3,4...

When only one end of a tube is closed, that end become a node and wave exhibits odd harmonics. For a tube with one close end and one open end:

$$f_n = \frac{nv}{4L}$$

where n = 1,3,5,7...

Wavelengths can be found by applying the formula f=v/λ.

Problem:
A string on a violin is 75 cm long. The speed of waves produced when the string is played is known to be 345 m/s. What is the fundamental frequency of the string?

Solution:
Remember that fundamental frequency is simply the first harmonic of the string. The wavelength of the first harmonic must be twice the length of the string:

$$\lambda = 2 \times l = 2 \times 0.75m = 1.5m$$

Then find the fundamental frequency:

$$f = \frac{v}{\lambda} = \frac{345\frac{m}{s}}{1.5m} = 230s^{-1} = 230Hz$$

Skill 16.3 Analyzing the relationship between sound and human perception of sound

The decibel scale is used to measure sound intensity. It originated in a unit known as the bel which is defined as the reduction in audio level over 1 mile of a telephone cable. Since the bel describes such a large variation in sound, it became more common to use the decibel, which is equal to 0.1 bel. A decibel value is related to the intensity of a sound by the following equation:

$$X_{dB} = 10 \log_{10}\left(\frac{X}{X_0}\right)$$

Where X_{dB} is the value of the sound in decibels
 X is the intensity of the sound
 X_0 is a reference value with the same units as X. X_0 is commonly taken to be the threshold of hearing at $10^{-12} W/m^2$.

It is important to note the logarithmic nature of the decibel scale and what this means for the relative intensity of sounds. The perception of the intensity of sound increases logarithmically, not linearly. Thus, an increase of 10 dB corresponds to an increase by one order of magnitude. For example, a sound that is 20 dB is not twice as loud as sound that is 10 dB; rather, it is 10 times as loud. A sound that is 30 dB will the 100 times as loud as the 10 dB sound.

Finally, let's equate the decibel scale with some familiar noises. Below are the decibel values of some common sounds.

Whispering voice: 20 dB   Cheering football stadium: 110 dB
Quiet office: 60 dB   Jet engine (100 feet away): 150 dB
Traffic: 70 dB   Space shuttle liftoff (100 feet away): 190 dB

Skill 16.4 Describing and applying the Doppler effect

The Doppler effect is the name given to the perceived change in frequency that occurs when the observer or source of a wave is moving. Specifically, the perceived frequency increases when a wave source and observer move toward each other and decreases when a source and observer move away from each other. Thus, the source and/or observer velocity must be factored in to the calculation of the perceived frequency.

The mathematical statement of this effect is:

$$f' = f_0 \left(\frac{v \pm v_o}{v \pm v_s} \right)$$

where f'= observed frequency
f_0= emitted frequency
v= the speed of the waves in the medium
v_s= the velocity of the source (positive in the direction away from the observer)
v_o= the velocity of the observer (positive in the direction towards the source)

Note that any motion that changes the perceived frequency of a wave will cause the Doppler effect to occur. Thus, the wave source, the observer position, or the medium through which the wave travels could possess a velocity that would alter the observed frequency of a wave. You may view animations of stationary and moving wave sources at the following URL:

http://www.kettering.edu/~drussell/Demos/doppler/doppler.html

So, let's consider two examples involving sirens and analyze what happens when either the source or the observer moves. First, imagine a person standing on the side of a road and a police car driving by with its siren blaring. As the car approaches, the velocity of the car will mean sound waves will "hit" the observer as the car comes closer and so the pitch of the sound will be high. As it passes, the pitch will slide down and continue to lower as the car moves away from the observer. This is because the sound waves will "spread out" as the source recedes. Now consider a stationary siren on the top of fire station and a person driving by that station. The same Doppler effect will be observed: the person would hear a high frequency sound as he approached the siren and this frequency would lower as he passed and continued to drive away from the fire station.

Problem: A sound source moves towards an observer at 8m/s while the observer moves towards the source at 9m/s. If the observer detects a sound of frequency 83kHz, what is the frequency of the sound emitted? The speed of sound in the medium is 343 m/s.

Solution: Using the Doppler equation, the emitted frequency f is given by

$$83 = f \frac{343 + 9}{343 - 8}; \Rightarrow f = 79 kHz$$

TEACHER CERTIFICATION STUDY GUIDE

COMPETENCY 17.0 UNDERSTAND ELECTROMAGNETIC WAVES AND THE ELECTROMAGNETIC SPECTRUM

Skill 17.1 Identify the connection between Maxwell's equations and the generation and propagation of electromagnetic waves

Electromagnetic waves traveling through a uniform material can be either absorbed or transmitted. Transmission and absorption are complementary phenomena. Reflection, an alternative possibility, may take place if there are non-uniformities in the material and if those non-uniformities are not "matched" (i.e., if they do not both have the same wave impedance).

Maxwell's equations, given here in differential form, provide a time-tested description of electromagnetic waves:

$$\nabla \times \mathbf{E}(\mathbf{r},t) = -\frac{\partial \mathbf{B}(\mathbf{r},t)}{\partial t} \qquad \nabla \times \mathbf{H}(\mathbf{r},t) = \frac{\partial \mathbf{D}(\mathbf{r},t)}{\partial t} + \mathbf{J}(\mathbf{r},t)$$

$$\nabla \cdot \mathbf{D}(\mathbf{r},t) = \rho(\mathbf{r},t) \qquad \nabla \cdot \mathbf{B}(\mathbf{r},t) = 0$$

The two curl equations may be combined, along with either of the divergence equations, to decouple the magnetic and electric fields and to form the wave equation. The wave equation can be expressed in terms of either the magnetic field (**H**) or the electric field (**E**). In this case, for simplicity, homogeneous, isotropic materials are assumed.

$$\nabla^2 \mathbf{H}(\mathbf{r},t) = \mu\varepsilon \frac{\partial^2 \mathbf{H}(\mathbf{r},t)}{\partial t^2} - \nabla \times \mathbf{J}(\mathbf{r},t) \qquad \nabla^2 \mathbf{E}(\mathbf{r},t) = \mu\varepsilon \frac{\partial^2 \mathbf{E}(\mathbf{r},t)}{\partial t^2} + \mu \frac{\partial \mathbf{J}(\mathbf{r},t)}{\partial t}$$

Absorption can usually be modeled as the result of the medium having a finite conductivity. Ideally, in classical electrodynamics, a perfect insulator is expected to transmit all light without any absorption. The presence of mobile charge carriers, such as free or weakly bound electrons, provides a means for absorption, where, for example, electrons acquire energy due to the presence of an electric field (and its corresponding magnetic field). The motion of the electrons may be impeded, resulting in conversion of some of the energy into heat.

In the wave equation, the current density **J** can be expressed as the product of the conductivity σ (a property of the material) and the electric field **E**. The wave equation may be further reduced to the following, where ψ is either **E** or **H**.

$$\nabla^2 \psi(\mathbf{r},t) = \left[\mu\varepsilon \frac{\partial^2}{\partial t^2} + \mu\sigma \frac{\partial}{\partial t}\right]\psi(\mathbf{r},t)$$

PHYSICS

This form of the wave equation includes the conductivity σ, and thus accounts for absorption in the material. In the frequency domain, the permittivity ε can be combined with the conductivity to form a generalized complex permittivity, ε_c.

$$\nabla^2 \psi(\mathbf{r}) = \omega^2 \mu \varepsilon_c \psi(\mathbf{r}) \qquad \varepsilon_c = \varepsilon - i\omega\sigma$$

For the simple case of plane waves, the general solution to the above equation is a set of weighted exponentials that represent traveling waves in opposite directions. The general form of the exponential is the following.

$$\psi = e^{i\omega\sqrt{\mu\varepsilon_c}\,r}$$

Since ε is complex, the argument of the exponential is likewise complex and may be expressed as the sum of a real and an imaginary number.

$$\psi = e^{\alpha r} e^{i\beta r}$$

The above expression has two factors: the exp(iβr) term expresses the traveling aspect of the wave; the exp(αr) term expresses the material absorption. The absorption coefficient, α, is generally negative to avoid unphysical wave amplification. In the case where σ = 0, or when the material is non-absorptive, α = 0, and the wave is transmitted without absorption. For materials with a non-zero conductivity, α is finite and there is some amount of absorption, all depending on the material parameters.

Thus, Maxwell's equations can be used to provide a classical explanation of absorption of waves by way of finite material conductivity. Absorption typically leads to an exponential decay of waves, as shown in the case of a simple plane wave. Full transmission occurs in purely (i.e., ideally) non-conductive materials, since no energy is converted to heat.

Skill 17.2   Demonstrating knowledge of radiometry and photometry

Radiometry and photometry both involve the measurement of electromagnetic waves. The most important difference between the two is the range of the EM spectrum which they cover, as described below. Additionally, photometry attempts to quantify light as it is perceived by the human eye.

Radiometry covers the entire optical radiation spectrum, wavelengths between 0.01 and 1000 micrometers. Radiometry measures light in terms of its absolute power. There are a large number of possible units, but among the most common are watts/m^2 and photons/sec-steradian. Radiometry is most commonly used in astronomy and is an important tool for Earth remote sensing (sensors placed in Earth's orbit, used to study conditions on our planet).

Photometry is limited to the visible spectrum, 360 and 830 nanometers. Photometry is unique not only in the range of wavelengths which it covers, but in that its measurements allow us to quantify light in terms of its *perceived* brightness to the human eye rather than its absolute power. It is important to understand that the human eye is more sensitive to certain wavelengths of light than to others. Photometry aims to account for this by weighting the power of each wavelength by a factor that represents the sensitivity of the eye at the corresponding wavelength. Both visual and physical photometries exist.

Visual photometry: Uses the eye as a comparison detector, allowing the direct use of the eye's spectral response.

Physical photometry: Two basic options exist for physical photometry. In the first, an optical radiation detector is built to mimic the spectral response of the eye. In the second, spectroradiometric readings are used to perform appropriate calculations and determine the proper weighting. The typical model of the eye's response to light as a function of wavelength is given by the luminosity function.

As with radiometry, there are many possible units for photometry; some of the most common are lumens, lux, and candelas.

It is interesting to compare the analogous quantities obtained from radiometry to photometry.

Photometry	Radiometry
Luminance	Radiance
Luminous flux	Radiant flux
Luminous intensity	Radiant intensity

We can further compare photometry and radiometry by considering an example case in which we have green and red light sources of equivalent radiance. The eye is far more sensitive to green light than red, so despite their equal radiance values, the green source will have a much higher luminance than the red.

Skill 17.3   Describing the electromagnetic spectrum in terms of wavelength, frequency, and energy

The electromagnetic spectrum is measured using frequency (f) in hertz or wavelength (λ) in meters. The frequency times the wavelength of every electromagnetic wave equals the speed of light (3.0×10^8 meters/second).

Roughly, the range of wavelengths of the electromagnetic spectrum is:

	f	λ
Radio waves	$10^5 - 10^{-1}$ hertz	$10^3 - 10^9$ meters
Microwaves	$3 \times 10^9 - 3 \times 10^{11}$ hertz	$10^{-3} - 10^{-1}$ meters
Infrared radiation	$3 \times 10^{11} - 4 \times 10^{14}$ hertz	$7 \times 10^{-7} - 10^{-3}$ meters
Visible light	$4 \times 10^{14} - 7.5 \times 10^{14}$ hertz	$4 \times 10^{-7} - 7 \times 10^{-7}$ meters
Ultraviolet radiation	$7.5 \times 10^{14} - 3 \times 10^{16}$ hertz	$10^{-8} - 4 \times 10^{-7}$ meters
X-Rays	$3 \times 10^{16} - 3 \times 10^{19}$ hertz	$10^{-11} - 10^{-8}$ meters
Gamma Rays	$> 3 \times 10^{19}$ hertz	$< 10^{-11}$ meters

Radio waves are used for transmitting data. Common examples are television, cell phones, and wireless computer networks. Microwaves are used to heat food and deliver Wi-Fi service. Infrared waves are utilized in night vision goggles. Visible light we are all familiar with as the human eye is most sensitive to this wavelength range. UV light causes sunburns and would be even more harmful if most of it were not captured in the Earth's ozone layer. X-rays aid us in the medical field and gamma rays are most useful in the field of astronomy.

Skill 17.4   Describing how the wave theory of light is applied to a variety of phenomena (i.e., interference, diffraction, and polarization)

Diffraction occurs when part of a wave front is obstructed. Diffraction and interference are essentially the same physical process. Diffraction refers to various phenomena associated with wave propagation such as the bending, spreading, and interference of waves emerging from an aperture. It occurs with any type of wave including sound waves, water waves, and electromagnetic waves such as light and radio waves.

Other forms of diffraction:

i) **Particle diffraction:** It is the diffraction of particles such as electrons, which is used as a powerful argument for quantum theory. It is possible to observe the diffraction of particles such as neutrons or electrons and hence we are able to infer the existence of wave particle duality.

ii) **Bragg diffraction:** This is diffraction from a multiple slits, and is similar to what occurs when waves are scattered from a periodic structure such as atoms in a crystal or rulings on a diffraction grating. Bragg diffraction is used in X-ray crystallography to deduce the structure of a crystal from the angles at which the X-rays are diffracted from it.

Here, we take a close look at important phenomena like single-slit diffraction, double-slit diffraction, diffraction grating, other forms of diffraction and lastly interference.

1. **Single-slit diffraction:** The simplest example of diffraction is single-slit diffraction in which the slit is narrow and a pattern of semi-circular ripples is formed after the wave passes through the slit.

2. **Double-slit diffraction:** These patterns are formed by the interference of light diffracting through two narrow slits.

3. **Diffraction grating:** Diffraction grating is a reflecting or transparent element whose optical properties are periodically modulated. In simple terms, diffraction gratings are fine parallel and equally spaced grooves or rulings on a material surface. When light is incident on a diffraction grating, light is reflected or transmitted in discrete directions, called diffraction orders. Because of their light dispersive properties, gratings are commonly used in monochromators and spectrophotometers. Gratings are usually designated by their groove density, expressed in grooves/millimeter. A fundamental property of gratings is that the angle of deviation of all but one of the diffracted beams depends on the wavelength of the incident light.

Interference is described as the superposition of two or more waves resulting in a new wave pattern. Interference is involved in Thomas Young's double slit experiment where two beams of light which are coherent with each other interfere to produce an interference pattern. Light from any source can be used to obtain interference patterns.

For example, Newton's rings can be produced with sun light. However, in general, white light is less suited for producing clear interference patterns as it is a mix of a full spectrum of colors. Sodium light is close to monochromatic and is thus more suitable for producing interference patterns. The most suitable is laser light as it is almost perfectly monochromatic.

Polarization is a property of transverse waves that describes the plane perpendicular to the direction of travel in which the oscillation occurs. In unpolarized light, the transverse oscillation occurs in all planes perpendicular to the direction of travel. Polarized light (created, for instance, by using polarizing filters that absorb light oscillating in other planes) oscillates in only a selected plane. An everyday example of polarization is found in polarized sunglasses which reduce glare.

Skill 17.5 Analyzing applications of double-slit interference, diffraction gratings, and interferometers

Please see **Skill 15.5**.

COMPETENCY 18.0 UNDERSTAND RAY OPTICS

Skill 18.1 Includes applying the laws of reflection, total internal reflection, and refraction

Snell's Law describes how light bends, or refracts, when traveling from one medium to the next. It is expressed as

$$n_1 \sin\theta_1 = n_2 \sin\theta_2$$

where n_i represents the index of refraction in medium i, and θ_i represents the angle the light makes with the normal in medium i.

Problem: The index of refraction for light traveling from air into an optical fiber is 1.44. (a) In which direction does the light bend? (b) What is the angle of refraction inside the fiber, if the angle of incidence on the end of the fiber is 22°?

Solution: (a) The light will bend toward the normal since it is traveling from a rarer region (lower n) to a denser region (higher n).

(b) Let air be medium 1 and the optical fiber be medium 2:

$$n_1 \sin\theta_1 = n_2 \sin\theta_2$$
$$(1.00)\sin 22° = (1.44)\sin\theta_2$$
$$\sin\theta_2 = \frac{1.00}{1.44}\sin 22° = (.6944)(.3746) = 0.260$$
$$\theta_2 = \sin^{-1}(0.260) = 15°$$

The angle of refraction inside the fiber is $15°$.

Reflection may occur whenever a wave travels from a medium of a given refractive index to another medium with a different index. A certain fraction of the light is reflected from the interface and the remainder is refracted. However, when the wave is moving from a dense medium into one less dense, that is the refractive index of the first is greater than the second, a critical angle exists which will create a phenomenon known as total internal reflection. In this situation all of the wave is reflected. The critical angle of incidence θ_c is the one for which the angle of reflection is 90 degrees. Thus, according to Snell's law

$$n_1 \sin\theta_c = n_2 \sin 90° \Rightarrow \theta_c = \sin^{-1}\frac{n_2}{n_1}$$

Snell's law may also be used to understand the phenomenon of dispersion since it relates the angle of refraction at the boundary between two media to the relative refractive indices. Since different frequency components of visible light have different indices of refraction in any medium other than vacuum, each component of a beam of white light is refracted at a different angle when it crosses a surface resulting in a separation of the colors.

Skill 18.2   Using ray diagrams with lenses and mirrors

Lenses
A lens is a device that causes electromagnetic radiation to converge or diverge. The most familiar lenses are made of glass or plastic and designed to concentrate or disperse visible light. Two of the most important parameters for a lens are its thickness and its focal length. Focal length is a measure of how strongly light is concentrated or dispersed by a lens. For a convex or converging lens, the focal length is the distance at which a beam of light will be focused to a single spot. Conversely, for a concave or diverging lens, the focal length is the distance to the point from which a beam appears to be diverging.

The images produced by lenses can be either virtual or real. A virtual image is one that is created by rays of light that appear to diverge from a certain point. Virtual images cannot be seen on a screen because the light rays do not actually meet at the point where the image is located. If an image and object appear on the same side of a converging lens, that image is defined as virtual. For virtual images, the image location will be negative and the magnification positive. Real images, on the other hand, are formed by light rays actually passing through the image. Thus, real images are visible on a screen. Real images created by a converging lens are inverted and have a positive image location and negative magnification.

Plane mirrors
Plane mirrors form virtual images. In other words, the image is formed behind the mirror where light does not actually reach. The image size is equal to the object size and object distance is equal to the image distance; i.e. the image is the same distance behind the mirror as the object is in front of the mirror. Another characteristic of plane mirrors is left-right reversal.

Example: Suppose you are standing in front of a mirror with your right hand raised. The image in the mirror will be raising its left hand.

Problem: If a cat creeps toward a mirror at a rate of 0.20 m/s, at what speed will the cat and the cat's image approach each other?

Solution: In one second, the cat will be 0.20 meters closer to the mirror. At the same time, the cat's image will be 0.20 meters closer to the cat. Therefore, the cat and its image are approaching each other at the speed of 0.40 m/s.

Problem: If an object that is two feet tall is placed in front of a plane mirror, how tall will the image of the object be?

Solution: The image of the object will have the same dimensions as the actual object, in this case, a height of two feet. This is because the magnification of an image in a plane mirror is 1.

Curved mirrors are usually sections of spheres. In a concave mirror the inside of the spherical surface is silvered while in a convex mirror it is the outside of the spherical surface that is silvered.

Terminology associated with spherical mirrors:

> Principal axis: The line joining the center of the sphere (of which we imagine the mirror is a section) to the center of the reflecting surface.
>
> Center of curvature: The center of the sphere of which the mirror is a section.
>
> Vertex: The point on the mirror where the principal axis meets the mirror or the geometric center of the mirror.
>
> Focal point: The point at which light rays traveling parallel to the principal axis will meet after reflection in a concave mirror. For a convex mirror, it is the point from which light rays traveling parallel to the principal axis will appear to diverge from after reflection. The focal point is midway between the center of curvature and the vertex.
>
> Focal length: The distance between the focal point and the vertex.
>
> Radius of curvature: The distance between the center of curvature and the vertex, i.e. the radius of the sphere of which the mirror is a section. The radius of curvature is twice the focal length

Image characteristics for concave mirrors:

- If the object is located beyond the center of curvature, the image will be real, inverted, smaller and located between the focal point and center of curvature.
- If the object is located at the center of curvature, the image will be real, inverted, of the same height and also located at the center of curvature.
- If the object is located between the center of curvature and focal point, the image will be real, inverted, larger and located beyond the center of curvature.
- If the object is located at the focal point no image is formed.
- If the object is located between the focal point and vertex, the image will be virtual, upright, larger and located on the opposite side of the mirror.
- If the object is located at infinity (very far away), the image is real, inverted, smaller and located at the focal point.
- For convex mirrors, the image is always virtual, upright, reduced in size and formed on the opposite side of the mirror.

Ray diagrams are a convenient way to visualize the propagation of waves and to perform reasonably accurate calculations of the effect of mirrors and lenses on light.

For mirrors, the focal point is either real (for concave mirrors) or virtual (for convex mirrors). In either case, the focal point is found by looking at the behavior of two parallel rays incident upon the mirror.

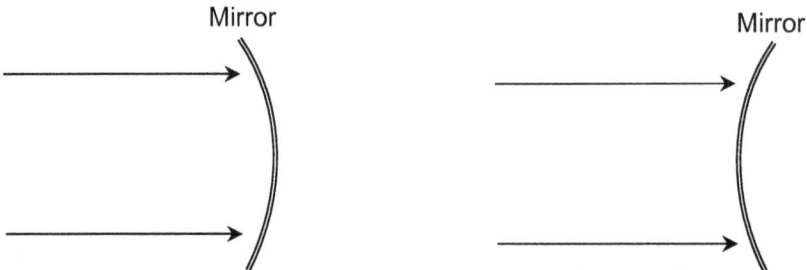

For each incident ray, the angle of reflection θ_r is equal to the angle of incidence θ_r, where the angle is measured from the normal to the mirror surface at the point of incidence.

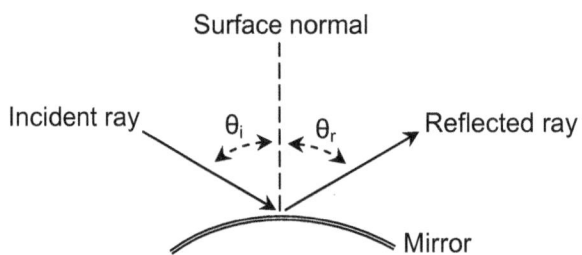

When this law is applied to the rays for either mirror, the focal points (f) are revealed as the (real or virtual) intersection of the rays. The virtual focus point is the intersection for the reflected rays when they are extended beyond the surface of the mirror.

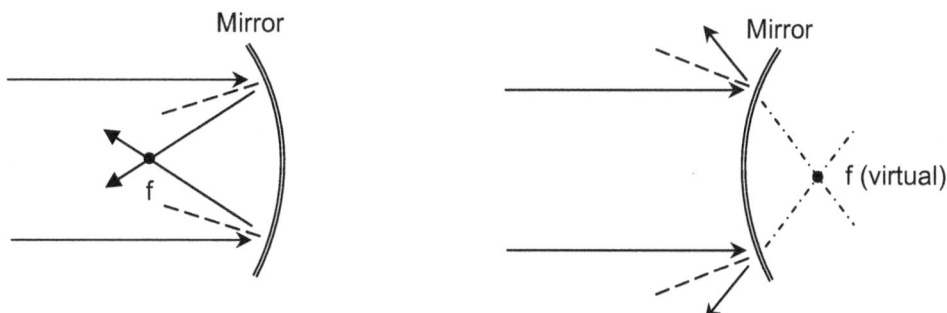

The focal point of a lens is found in a similar manner, with the exception that instead of using the law of reflection, refraction by way of Snell's law must be applied. Since real lenses have a finite thickness, Snell's law must be used for the ray both as it enters the lens material and as it exits the lens material.

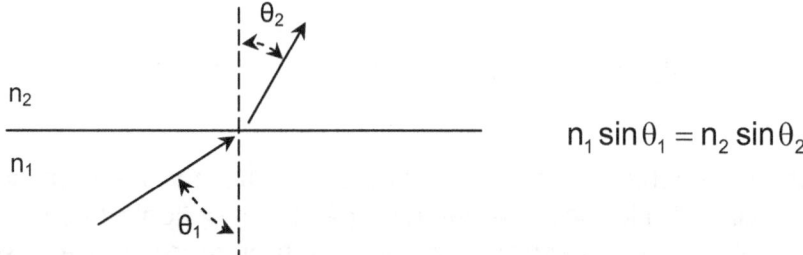

$$n_1 \sin \theta_1 = n_2 \sin \theta_2$$

As with reflection, refraction requires consideration of the normal to the surface. Also, the refractive indices must be used. It is assumed here that the refractive index of the outer medium is n_1 and the refractive index of the lens is n_2, and that $n_1 < n_2$. The intersection of parallel incident rays determines the focal point, which may again be either real or virtual. Real focal points occur on the opposite side of the lens from the source of illumination, and virtual focal points occur on the same side as the source of illumination.

Only one lens case is shown here, but the principles apply equally to all variations of lenses.

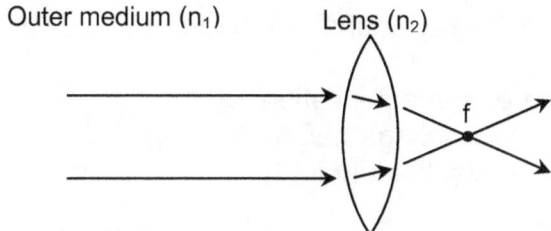

Care must be taken in properly identifying the normal and in applying Snell's law to both incidences of refraction for each ray.

Determining the point of image formation for mirrors and lenses is accomplished in a similar manner to that of the focal points. As with the focal points, image points may be either real or virtual, depending on the characteristics of the mirror or lens. To find the image point, two rays of differing angles must be traced as they interact with the lens or mirror. The initial directions of the rays can be chosen arbitrarily, but it is ideal to choose the directions such that the difficulty with determining the direction of the reflected or refracted ray is minimized. The intersection of these two rays is the image point. Only two examples are shown here, but the principles behind these examples may be applied to any variation of the situations, as well as to any combination of lenses and mirrors.

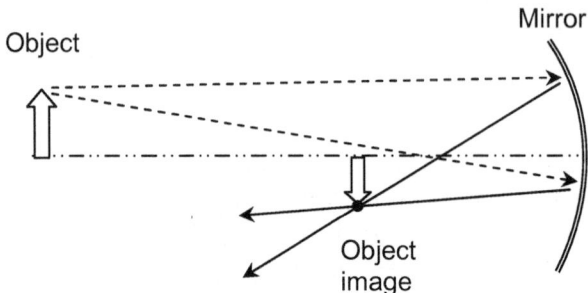

Skill 18.3 Applying the thin lens and spherical mirror equations

Lenses
A thin lens in one in which focal length is much greater than lens thickness. For problems involving thin lenses, we can disregard any optical effects of the lens itself. Additionally, we can assume that the light that interacts with the lens makes a small angle with the optical axis of the system and so the sine and tangent values of the angle are approximately equal to the angle itself. This paraxial approximation, along with the thin lens assumptions, allows us to state:

$$\frac{1}{s} + \frac{1}{s'} = \frac{1}{f}$$

Where s = distance from the lens to the object (object location)
s' = distance from the lens to the image (image location)
f = focal length of the lens

Most lenses also cause some magnification of the object. Magnification is defined as:

Where m = magnification
y' = image height
y = object height

$$m = \frac{y'}{y} = -\frac{s'}{s}$$

Sign conventions will make it easier to understand thin lens problems:

Focal length: positive for a converging lens; negative for a diverging lens
Object location: positive when in front of the lens; negative when behind the lens
Image location: positive when behind the lens; negative when in front of the lens
Image height: positive when upright; negative when upside-down.
Magnification: positive for an erect, virtual image; negative for an inverted, real image

Problem:
A converging lens has a focal length of 10.00 cm and forms a 2.0 cm tall image of a 4.00 mm tall real object to the left of the lens. If the image is erect, is the image real or virtual? What are the locations of the object and the image?

Solution:
We begin by determining magnification:
$$m = \frac{y'}{y} = \frac{0.02m}{0.004m} = 5$$

Since the magnification is positive and the image is erect, we know the image must be virtual.

To find the locations of the object and image, we first relate them by using the magnification:
$$m = -\frac{s'}{s}$$
$$s' = -ms$$

Then we substitute into the thin lens equation, creating one variable in one unknown:
$$\frac{1}{s} + \frac{1}{s'} = \frac{1}{f}$$
$$\frac{1}{s} - \frac{1}{5s} = \frac{1}{10cm}$$
$$\frac{5-1}{5s} = \frac{1}{10cm}$$
$$s = \frac{40cm}{5} = 8cm \rightarrow\rightarrow s' = -5 \times 8cm = -40cm$$

Thus the object is located 8 cm to the left of the lens and the image is 40 cm to the left of the lens.

Mirrors

The relationship between the object distance from vertex s, the image distance from vertex s', and the focal length f is given by the equation

$$\frac{1}{s} + \frac{1}{s'} = \frac{1}{f}$$

Magnification is defined as:

$$m = \frac{y'}{y} = -\frac{s'}{s}$$

where m=magnification, y'=image height, y=object height

Problem: A concave mirror collects light from a star. If the light rays converge at 50 cm, what is the radius of curvature of the mirror?

Solution: The point at which the rays converge is known as the focal point. The focal length, in this case, 50 cm, is the distance from the focal point to the mirror. The radius of curvature is the distance from the vertex to the center of curvature. The vertex is the point on the mirror where the principal axis meets the mirror. The center of curvature represents the point in the center of the sphere from which the mirror was sliced. Since the focal point is the midpoint of the line from the vertex to the center of curvature, or focal length, the focal length would be one-half the radius of curvature. Since the focal length in this case is 50 cm, the radius of curvature would be 100 cm.

Problem: An image of an object in a mirror is upright and reduced in size. In what type of mirror is this image being viewed, plane, concave, or convex?

Solution: The image in a plane mirror would be the same size as the object. The image in a concave mirror would be magnified if upright. Only a convex mirror would produce a reduced upright image of an object.

Skill 18.4 Explaining the operation of optical instruments

Eye

The eye is a very complex sensory organ. Although there are many critical anatomical features of the eye, the lens and retina are the most important for focusing and sensing light. Light passes through the cornea and through the lens. The lens is attached to numerous muscles that contract and relax to move the lens in order to focus the light onto the retina. The pupil also contracts and relaxes to allow more or less light in the eye as required.

The retina contains rod cells which are responsible for vision in low light and cone cells which sense color and detail. Different types of cone cells are capable of sensing different wavelengths of light. A chemical called rhodopsin is present in the retina that converts light signals into electrical impulses that are sent to the brain to interpret as vision. The retina is lined with a black pigment called melanin that reduces reflection.

The eye relies on refraction to focus light onto the retina. Refraction occurs at four curved interfaces; between the air and the front of the cornea, the back of the cornea and the aqueous humor, the aqueous humor at the front of the lens, and the back of the lens and the vitreous humor. When each of these interfaces are working properly the light arrives at the retina in perfect focus for transmission to the brain as an image.

Eye Glasses

When all the parts of the eye are not working together correctly, corrective lenses or eyeglasses may be needed to assist the eye in focusing the light onto the retina. The surfaces of the lens or cornea may not be smooth causing the light to refract in the wrong direction. This is called astigmatism. Another common problem is that the lens is not able to change its curvature appropriately to match the image. The cornea can also be misshaped resulting in blurred vision. Corrective lenses consist of curves pieces of glass which bend the light in order to change the focal point of the light. A nearsighted eye forms images in front of the retina. To correct this, a minus lens consisting of two concave prisms is used to bend light out and move the image back to the retina. A farsighted eye creates images behind the retina. This is corrected using plus lenses that bend light in and bring the image forward onto the retina. The worse the vision, the farther out of focus the image is on the retina. Therefore the stronger the lens the further the focal point is moved to compensate.

Spectroscope

Spectrometers known as spectroscopes are used to identify materials. Spectroscopes are used often in astronomy and some branches of chemistry. Early spectroscopes were simply a prism with graduations marking wavelengths of light. Modern spectroscopes typically use a diffraction grating, a movable slit, and some kind of photo detector, all automated and controlled by a computer.

When materials are heated they emit light that is characteristic of its atomic composition. The emission of certain frequencies of light produce a pattern of lines that are comparable to a fingerprint. The yellow light emission of heated sodium is a typical example.

A spectroscope is able to detect, measure and record the frequencies of the emitted light. This is done by passing the light though a slit to a collimating lens which transforms the light into parallel rays. The light is then passed through a prism that refracts the beam into a spectrum of different wavelengths. The image is then viewed alongside a scale to determine the characteristic wavelengths. Spectral analysis is an important tool for determining and analyzing the composition of unknown materials as well as for astronomical studies.

Camera

A camera is another device that utilizes the lens' ability to refract light to capture and process an image. As with the eye, light enters the lens of a camera and focuses the light on the other side. Instead of focusing on the retina, the image is focused on the film to create a film negative. This film negative is later processed with chemicals to create a photograph. A camera uses a converging or convex lens. This lens captures and directs light to a single point to create a real image on the surface of the film. To focus a camera on an image, the distance of the lens from the film is adjusted in order to ensure that the real image converges on the surface of the film and not in front of or behind it.

Different lenses are available which capture and bend the light to different degrees. A lens with more pronounced curvature will be able to bend the light more acutely causing the image to converge more closely to the lens. Conversely a flatter lens will have a longer focal distance. The further the lens is located from the film (flatter lens), the larger the image becomes. Thus zoom lenses on cameras are flat while wide angle lenses are more rounded. The focal length number on a certain lens conveys the magnification ability of the lens.

The film functions like the retina of the eye in that it is light sensitive and can capture light images when exposed. However, this exposure must be brief to capture the contrasting amounts of light and a clear image. The rest of the camera functions to precisely control how much light contacts the film. The aperture is the lens opening which can open and close to let in more or less light. The temporal length of light exposure is controlled by the shutter which can be set at different speeds depending on the amount of action and level of light available. The film speed refers to the size of the light sensitive grains on the surface of the film. The larger grains absorb more light photons than the smaller grains, so film speed should be selected according to lighting conditions.

Telescope

A telescope is a device that has the ability to make distant objects appear to be much closer. Most telescopes are one of two varieties, a refractor which uses lenses or a reflector which uses mirrors. Each accomplishes the same purpose but in totally different ways. The basic idea of a telescope is to collect as much light as possible, focus it, and then magnify it. The objective lens or primary mirror of a telescope brings the light from an object into focus. An eyepiece lens takes the focused light and "spreads it out" or magnifies it using the same principle as a magnifying glass using two curved surfaces to refract the light.

Microscope

Microscopes are used to view objects that are too small to be seen with the naked eye. A microscope usually has an objective lens that collects light from the sample and an eyepiece which brings the image into focus for the observer. It also has a light source to illuminate the sample. Typical optical microscopes achieve magnification of up to 1500 times.

Skill 18.5 Describing the effect of limit resolution

When physicists discuss resolution in an optical system, it usually refers to how close two objects can be and still be recognized as distinct. The following diagram demonstrates that as two objects become closer together, it becomes impossible to note that there are two separate objects. The minimum distance at which two objects can be resolved is given by the Rayleigh criterion.

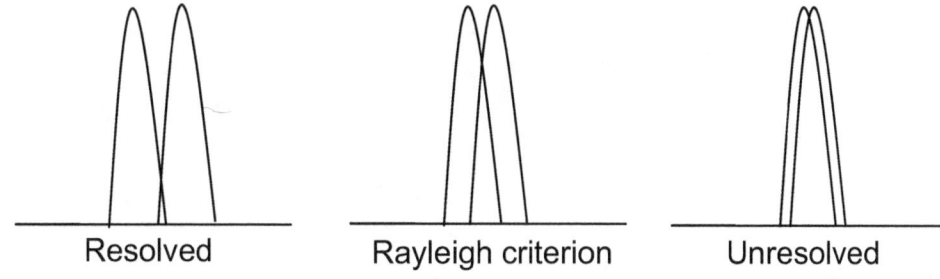

Resolved   Rayleigh criterion   Unresolved

The resolution that can be obtained by a given optical system (anything from a camera to high power microscope to the human eye) depends upon the size of the aperture and wavelength of the light (or other EM wave) involved. We can state the Rayleigh criterion mathematically by defining angular resolution, which is the resolving power of a given image forming device. The following equation is for a circular aperture:

$$\sin\theta = 1.22\frac{\lambda}{D}$$

Where θ=angular resolution
λ=wavelength
D=the diameter of the aperture

Thus, higher resolution can be obtained with shorter wavelengths. This concept is important in both human vision and in imaging systems. It explains, for instance, why electron microscopy allows for a higher level of resolution than light microscopy. The electron beam has a much shorter wavelength than visible light.

SUBAREA V. NATURE OF MATTER, THERMODYNAMICS, AND MODERN PHYSICS

COMPETENCY 19.0 UNDERSTAND THE PARTICULATE NATURE OF MATTER

Skill 19.1 Includes recognizing basic characteristics of the states of matter

The phase or state of matter (solid, liquid, or gas) is identified by its shape and volume. A solid has a definite shape and volume. A liquid has a definite volume, but no shape. A gas has no shape or volume because it will spread out to occupy the entire space of whatever container it is in. While plasma is really a type of gas, its properties are so unique that it is considered a unique phase of matter. Plasma is a gas that has been ionized; meaning that at least on electron has been removed from some of its atoms. Plasma shares some characteristics with gas, specifically, the high kinetic energy of its molecules. Thus, plasma exists as a diffuse "cloud," though it sometimes includes tiny grains (this is termed dusty plasma). What most distinguishes plasma from gas is that it is electrically conductive and exhibits a strong response to electromagnetic fields. This property is a consequence of the charged particles that result from the removal of electrons from the molecules in the plasma.

Molecules have kinetic energy (they move around), and they also have intermolecular attractive forces (they stick to each other). The relationship between these two determines whether a collection of molecules will be a gas, liquid, or solid.

A gas has an indefinite shape and an indefinite volume. The kinetic model for a gas is a collection of widely separated molecules, each moving in a random and free fashion, with negligible attractive or repulsive forces between them. Gases will expand to occupy a larger container so there is more space between the molecules. Gases can also be compressed to fit into a small container so the molecules are less separated. Diffusion occurs when one material spreads into or through another. Gases diffuse rapidly and move from one place to another.

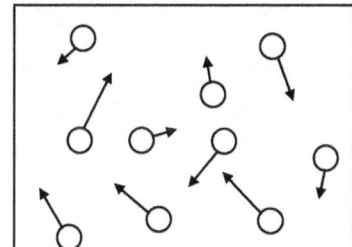

A liquid assumes the shape of the portion of any container that it occupies and has a specific volume. The kinetic model for a liquid is a collection of molecules attracted to each other with sufficient strength to keep them close to each other but with insufficient strength to prevent them from moving around randomly. Liquids have a higher density and are much less compressible than gases because the molecules in a liquid are closer together.

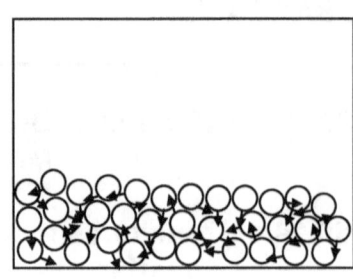

Diffusion occurs more slowly in liquids than in gases because the molecules in a liquid stick to each other and are not completely free to move.

A solid has a definite volume and definite shape. The kinetic model for a solid is a collection of molecules attracted to each other with sufficient strength to essentially lock them in place. Each molecule may vibrate, but it has an average position relative to its neighbors. If these positions form an ordered pattern, the solid is called crystalline. Otherwise, it is called amorphous. Solids have a high density and are almost incompressible because the molecules are close together. Diffusion occurs extremely slowly because the molecules almost never alter their position.

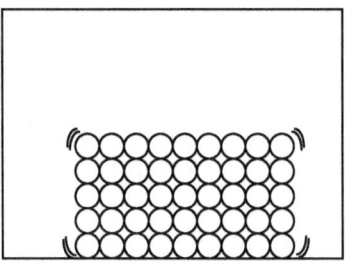

Phase changes occur when the relative importance of kinetic energy and intermolecular forces is altered sufficiently for a substance to change its state. The transition from gas to liquid is called condensation and from liquid to gas is called vaporization. The transition from liquid to solid is called freezing and from solid to liquid is called melting. The transition from gas to solid is called deposition and from solid to gas is called sublimation.

Heat removed from a substance during condensation, freezing, or deposition permits new intermolecular bonds to form, and heat added to a substance during vaporization, melting, or sublimation breaks intermolecular bonds.

During these phase transitions, this latent heat is removed or added with no change in the temperature of the substance because the heat is not being used to alter the speed of the molecules or the kinetic energy when they strike each other or the container walls. Latent heat alters intermolecular bonds.

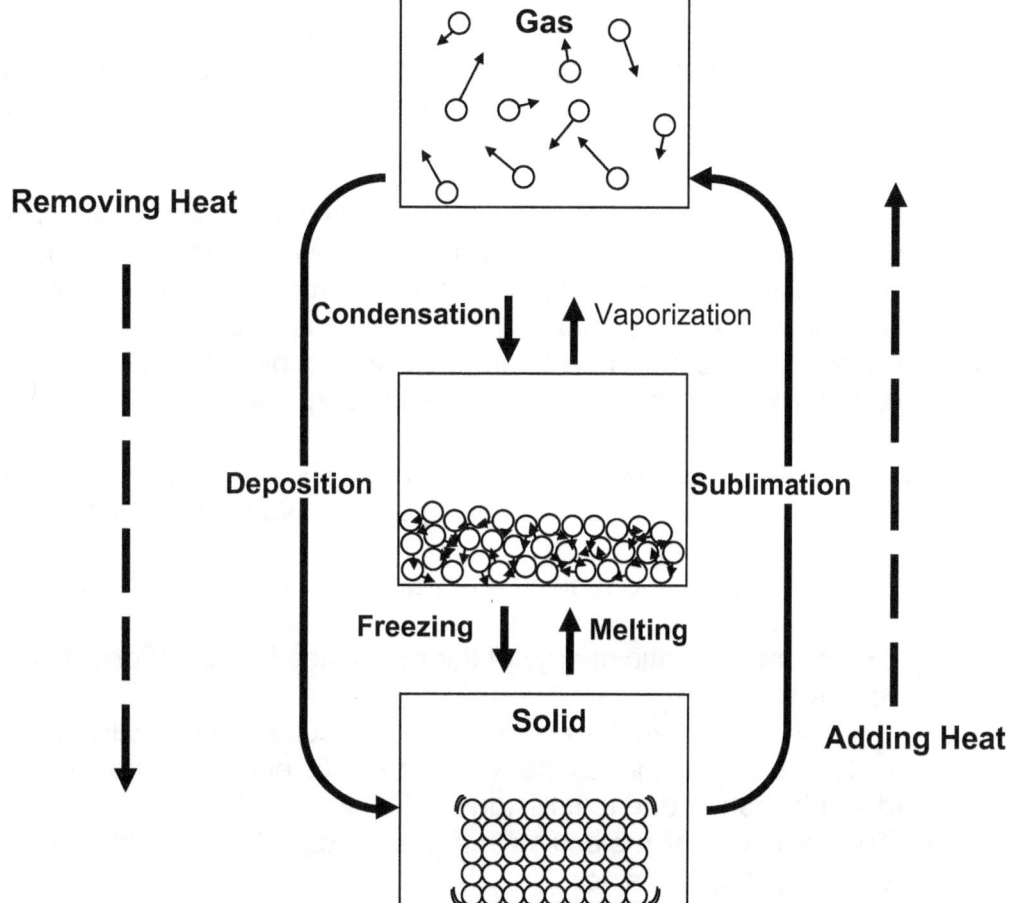

Skill 19.2 Describing how the Maxwell-Boltzmann theory applies to an ideal gas

The relationship between kinetic energy and intermolecular forces determines whether a collection of molecules will be a gas, liquid, or solid. In a gas, the energy of intermolecular forces is much weaker than the kinetic energy of the molecules. Kinetic molecular theory is usually applied to gases and is best applied by imagining ourselves shrinking down to become a molecule and picturing what happens when we bump into other molecules and into container walls.

Gas pressure results from molecular collisions with container walls. The number of molecules striking an area on the walls and the average kinetic energy per molecule are the only factors that contribute to pressure. A higher temperature increases speed and kinetic energy. There are more collisions at higher temperatures, but the average distance between molecules does not change, and thus density does not change in a sealed container.

Kinetic molecular theory explains why the pressure and temperature of ideal gases behave the way they do by making a few assumptions, namely:

1) The energies of intermolecular attractive and repulsive forces may be neglected.
2) The average kinetic energy of the molecules is proportional to absolute temperature.
3) Energy can be transferred between molecules during collisions and the collisions are elastic, so the average kinetic energy of the molecules doesn't change due to collisions.
4) The volume of all molecules in a gas is negligible compared to the total volume of the container.

Strictly speaking, molecules also contain some kinetic energy by rotating or experiencing other motions. The motion of a molecule from one place to another is called translation. Translational kinetic energy is the form that is transferred by collisions, and kinetic molecular theory ignores other forms of kinetic energy because they are not proportional to temperature.

The following table summarizes the application of kinetic molecular theory to an increase in container volume, number of molecules, and temperature:

Effect of an **increase** in one variable with other two constant	Impact on gas: − = decrease, 0 = no change, + = increase						
	Average distance between molecules	Density in a sealed container	Average speed of molecules	Average translational kinetic energy of molecules	Collisions with container walls per second	Collisions per unit area of wall per second	Pressure (P)
Volume of container (V)	+	−	0	0	−	−	−
Number of molecules	−	+	0	0	+	+	+
Temperature (T)	0	0	+	+	+	+	+

Additional details on the kinetic molecular theory may be found at http://hyperphysics.phy-astr.gsu.edu/hbase/kinetic/ktcon.html. An animation of gas particles colliding is located at http://comp.uark.edu/~jgeabana/mol_dyn/.

The pressure, temperature and volume relationships for an ideal gas (a gas described by the assumptions of the kinetic molecular theory listed above) are given by the following gas laws:

Boyle's law states that the volume of a fixed amount of gas at constant temperature is inversely proportional to the gas pressure, or:

$$V \propto \frac{1}{P}.$$

Gay-Lussac's law states that the pressure of a fixed amount of gas in a fixed volume is proportional to absolute temperature, or:

$$P \propto T.$$

Charles's law states that the volume of a fixed amount of gas at constant pressure is directly proportional to absolute temperature, or:

$$V \propto T.$$

The **combined gas law** uses the above laws to determine a proportionality expression that is used for a constant quantity of gas:

$$V \propto \frac{T}{P}.$$

The combined gas law is often expressed as an equality between identical amounts of an ideal gas at two different states ($n_1=n_2$):

$$\frac{P_1V_1}{T_1} = \frac{P_2V_2}{T_2}.$$

Avogadro's hypothesis states that equal volumes of different gases at the same temperature and pressure contain equal numbers of molecules. **Avogadro's law** states that the volume of a gas at constant temperature and pressure is directly proportional to the quantity of gas, or:

$V \propto n$ where n is the number of moles of gas.

Avogadro's law and the combined gas law yield $V \propto \frac{nT}{P}$. The proportionality constant R—the **ideal gas constant**—is used to express this proportionality as the **ideal gas law**: $PV = nRT$.

The ideal gas law is useful because it contains all the information of Charles's, Avogadro's, Boyle's, and the combined gas laws in a single expression.

Solving ideal gas law problems is a straightforward process of algebraic manipulation. **Errors commonly arise from using improper units**, particularly for the ideal gas constant R. An absolute temperature scale must be used—never °C—and is usually reported using the Kelvin scale, but volume and pressure units often vary from problem to problem.

If pressure is given in atmospheres and volume is given in liters, a value for R of **0.08206 L-atm/(mol-K)** is used. If pressure is given in pascal (newtons/m^2) and volume in m^3, then the SI value for R of **8.314 J/(mol-K)** may be used because a joule is defined as a newton-meter or a pascal-m^3. A value for R of **8.314 Pa-m^3/(mol-K)** is identical to the ideal gas constant using joules.

The ideal gas law may also be rearranged to determine gas molar density in moles per unit volume (molarity):

$$\frac{n}{V} = \frac{P}{RT}.$$

Gas density d in grams per unit volume is found after multiplication by the molecular weight M:

$$d = \frac{nM}{V} = \frac{PM}{RT}.$$

Molecular weight may also be determined from the density of an ideal gas:

$$M = \frac{dV}{n} = \frac{dRT}{P}.$$

Example: Determine the molecular weight of an ideal gas that has a density of 3.24 g/L at 800 K and 3.00 atm.

Solution: $M = \dfrac{dRT}{P} = \dfrac{\left(3.24\ \frac{g}{L}\right)\left(0.08206\ \frac{L\text{-atm}}{mol\text{-}K}\right)(800\ K)}{3.00\ atm} = 70.9\ \dfrac{g}{mol}$.

Tutorials for gas laws may be found online at:
http://www.chemistrycoach.com/tutorials-6.htm.

A flash animation tutorial for problems involving a piston may be found at http://www.mhhe.com/physsci/chemistry/essentialchemistry/flash/gasesv6.swf.

Skill 19.3 Analyzing Phase Changes

A substance's molar heat capacity is the heat required to change the temperature of one mole of the substance by one degree. Heat capacity has units of joules per mol- kelvin or joules per mol- °C. The two units are interchangeable because we are only concerned with differences between one temperature and another. A Kelvin degree and a Celsius degree are the same size.

The specific heat of a substance (also called specific heat capacity) is the heat required to change the temperature of one gram or kilogram by one degree. Specific heat has units of joules per gram-°C or joules per kilogram-°C.

The temperature of a material rises when heat is transferred to it and falls when heat is removed from it. When the material is undergoing a phase change (e.g. from solid to liquid), however, it absorbs or releases heat without a corresponding change in temperature. The heat that is absorbed or released during phase change is known as latent heat.

A temperature vs. heat graph can demonstrate these relationships visually. One can also calculate the specific heat or latent heat of phase change for the material by studying the details of the graph.

Example: The plot below shows heat applied to 1g of ice at -40C. The horizontal parts of the graph show the phase changes where the material absorbs heat but stays at the same temperature. The graph shows that ice melts into water at 0C and the water undergoes a further phase change into steam at 100C.

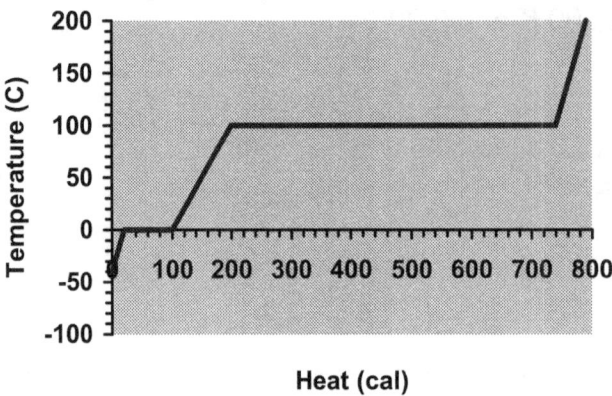

The specific heat of ice, water and steam and the latent heat of fusion and vaporization may be calculated from each of the five segments of the graph.

For instance, we see from the flat segment at temperature 0C that the ice absorbs 80 cal of heat. The latent heat L of a material is defined by the equation $\Delta Q = mL$ where ΔQ is the quantity of heat transferred and m is the mass of the material. Since the mass of the material in this example is 1g, the latent heat of fusion of ice is given by $L = \Delta Q / m$ = 80 cal/g.

The next segment shows a rise in the temperature of water and may be used to calculate the specific heat C of water defined by $\Delta Q = mC\Delta T$, where ΔQ is the quantity of heat absorbed, m is the mass of the material and ΔT is the change in temperature. According to the graph, ΔQ = 200-100 =100 cal and ΔT = 100-0=100C. Thus, C = 100/100 = 1 cal/gC.

Problem: The plot below shows the change in temperature when heat is transferred to 0.5g of a material. Find the initial specific heat of the material and the latent heat of phase change.

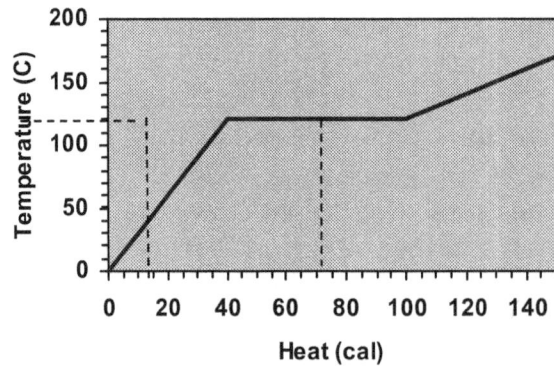

Solution:
Looking at the first segment of the graph, we see that ΔQ = 40 cal and ΔT = 120 C. Since the mass m = 0.5g, the specific heat of the material is given by
$C = \Delta Q / (m \Delta T)$ = 40/(0.5 X120) = 0.67 cal/gC.

The flat segment of the graph represents the phase change. Here ΔQ = 100 - 40=60 cal. Thus, the latent heat of phase change is given by $L = \Delta Q / m$ = 60/(0.5) = 120 cal/g.

Skill 19.4   Describing the properties of materials at low temperatures

Conductors are those materials which allow for the free passage of electrical current. However, all materials exhibit a certain opposition to the movement of electrons. This opposition is known as resistivity (ρ). Resistivity is determined experimentally by measuring the resistance of a uniformly shaped sample of the material and applying the following equation:

$$\rho = R \frac{A}{l}$$

where ρ = static resistivity of the material
R = electrical resistance of the material sample
A = cross-sectional area of the material sample
L = length of the material sample

The temperature at which these measurements are taken is important as it has been shown that resistivity is a function of temperature. For conductors, resistivity increases with increasing temperature and decreases with decreasing temperature. At extremely low temperatures resistivity assumes a low and constant value known as residual resistivity (ρ_0). Residual resistivity is a function of the type and purity of the conductor.

The following equation allows us to calculate the resistivity ρ of a material at any temperature given the resistivity at a reference temperature, in this case at $20°C$:

$$\rho = \rho_{20}[1 + a(t - 20)]$$

where ρ_{20} = resistivity at $20°C$
a= proportionality constant characteristic of the material
t=temperature in Celsius

PHYSICS

Problem: The tungsten filament in a certain light bulb is a wire 8 μm in diameter and 10 mm long. Given that, for tungsten, $\rho_{20} = 5.5 \times 10^{-8}$ Ω·m and $a = 4.5 \times 10^{-3} K^{-1}$, what will the resistance of the filament be at 45°C?

Solution: First we must find the resistivity of the tungsten at 45°C:

$$\rho = 5.5 \times 10^{-8}(1 + 4.5 \times 10^{-3}(45 - 20)) = 6.1 \times 10^{-8} \Omega \cdot m$$

Now we can rearrange the equation defining resistivity and solve for the resistance of the filament: $R = \rho \dfrac{l}{A} = 6.1 \times 10^{-8} \times 0.01 / (\pi(4 \times 10^{-6})^2) = 12.1 \Omega$

TEACHER CERTIFICATION STUDY GUIDE

COMPETENCY 20.0 UNDERSTAND THE LAWS OF THERMODYNAMICS

Skill 20.1   Includes differentiating between temperature, internal energy, and heat

Heat is generally measured in terms of temperature, a measure of the average internal energy of a material. Temperature is an intensive property, meaning that it does not depend on the amount of material. Heat content is an extensive property because more material at the same temperature will contain more heat. The relationship between the change in heat energy of a material and the change in its temperature is given by $\Delta Q = mC\Delta T$, where ΔQ is the change in heat energy, m is the mass of the material, ΔT is the change in temperature and C is the specific heat which is characteristic of a particular material.

The zeroth law of thermodynamics generally deals with bodies in thermal equilibrium with each other and is the basis for the idea of temperature. Most commonly, the law is stated as, "If two thermodynamic systems are in thermal equilibrium with a third, they are also in thermal equilibrium with each other." Said another way, this very basic law simply states that if object A is same temperature as object B, and object C is the same temperature as object B, then object A and C are also the same temperature.

Skill 20.2   Calculating heat loss or gain using specific heat

The specific heat of a substance (also called specific heat capacity) is the heat required to change the temperature of one gram or kilogram by one degree. Specific heat has units of joules per gram-°C or joules per kilogram-°C.

These terms are used to solve problems involving a change in temperature by applying the formula: $q = n \times C \times \Delta T$

> Where q → heat added (positive) or evolved (negative)
> n → amount of material
> C → molar heat capacity if n is in moles, specific heat if is a mass
> ΔT → change in temperature $T_{final} - T_{initial}$

<u>Example:</u>
What is the change in energy of 10 g of gold at 25 °C when it is heated beyond its melting point to 1300 °C. You will need the following data for gold:

Solid heat capacity = 28 J/mol-K   Molten heat capacity: 20 J/mol-K
Enthalpy of fusion = 12.6 kJ/mol   Melting Point 1064°C

Solution: First determine the number of moles used: $10 \text{ g} \times \dfrac{1 \text{ mol}}{197 \text{ g}} = 0.051 \text{ mol}$.

There are then three steps:
1) Heat the solid
2) Melt the solid
3) Heat the liquid

All three require energy so they will be positive numbers.
   1) Heat the solid:

$$q_1 = n \times C \times \Delta T = 0.051 \text{ mol} \times 28 \dfrac{\text{J}}{\text{mol-K}} \times (1064 \text{ °C} - 25 \text{ °C})$$

$$= 1.48 \times 10^3 \text{ J} = 1.48 \text{ kJ}$$

   2) Melt the solid: $q_2 = n \times \Delta H_{fusion} = 0.051 \text{ mol} \times 12.6 \dfrac{\text{kJ}}{\text{mol}}$

$$= 0.64 \text{ kJ}$$

   3) Heat the liquid:

$$q_3 = n \times C \times \Delta T = 0.051 \text{ mol} \times 20 \dfrac{\text{J}}{\text{mol-K}} \times (1300 \text{ °C} - 1064 \text{ °C})$$

$$= 2.4 \times 10^2 \text{ J} = 0.24 \text{ kJ}$$

The sum of the three processes is the total change in energy of the gold:

$$q = q_1 + q_2 + q_3 = 1.48 \text{ kJ} + 0.64 \text{ kJ} + 0.24 \text{ kJ} = 2.36 \text{ kJ}$$
$$= 2.4 \text{ kJ}$$

Skill 20.3 Identifying processes of thermal energy transfer

All heat transfer is the movement of thermal energy from hot to cold matter. This movement down a thermal gradient is a consequence of the second law of thermodynamics. The three methods of heat transfer are listed and explained below.

Conduction: Electron diffusion or photo vibration is responsible for this mode of heat transfer. The bodies of matter themselves do not move; the heat is transferred because adjacent atoms that vibrate against each other or as electrons flow between atoms. This type of heat transfer is most common when two solids come in direct contact with each other. This is because molecules in a solid are in close contact with one another and so the electrons can flow freely. It stands to reason, then, that metals are good conductors of thermal energy. This is because their metallic bonds allow the freest movement of electrons. Similarly, conduction is better in denser solids. Examples of conduction can be seen in the use of copper to quickly convey heat in cooking pots, the flow of heat from a hot water bottle to a person's body, or the cooling of a warm drink with ice.

Convection: Convection involves some conduction but is distinct in that it involves the movement of warm particles to cooler areas. Convection may be either natural or forced, depending on how the current of warm particles develops. Natural convection occurs when molecules near a heat source absorb thermal energy (typically via conduction), become less dense, and rise. Cooler molecules then take their place and a natural current is formed. Forced convection, as the name suggests, occurs when liquids or gases are moved by pumps, fans, or other means to be brought into contact with warmer or cooler masses. Because the free motion of particles with different thermal energy is key to this mode of heat transfer, convection is most common in liquid and gases. Convection can, however, transfer heat between a liquid or gas and a solid. Forced convection is used in "forced air" home heating systems and is common in industrial manufacturing processes. Additionally, natural convection is responsible for ocean currents and many atmospheric events. Finally, natural convection often arises in association with conduction, for instance in the air near a radiator or the water in a pot on the stove.

Radiation: This method of heat transfer occurs via electromagnetic radiation. All matter warmer than absolute zero (that is, all known matter) radiates heat. This radiation occurs regardless of the presence of any medium. Thus, it occurs even in a vacuum. Since light and radiant heat are both part of the EM spectrum, we can easily visualize how heat is transferred via radiation. For instance, just like light, radiant heat is reflected by shiny materials and absorbed by dark materials. Common examples of radiant heat include the way sunlight travels from the sun to warm the earth, the use of radiators in homes, and the warmth of incandescent light bulbs.

Skill 20.4 Applying the principles of enthalpy, internal energy, and thermodynamic work

The internal energy of a material is the sum of the total kinetic energy of its molecules and the potential energy of interactions between those molecules. Total kinetic energy includes the contributions from translational motion and other components of motion such as rotation. The potential energy includes energy stored in the form of resisting intermolecular attractions between molecules.

The enthalpy (H) of a material is the sum of its internal energy and the mechanical work it can do by driving a piston. A change in the enthalpy of a substance is the total energy change caused by adding/removing heat at constant pressure.

The First Law of Thermodynamics:
The first law of thermodynamics is a general expression of the conservation of energy wherein heat transfer is considered to be a form of energy transfer in addition to work done on or by a system. The internal energy U of a closed system is related to the heat energy Q in the system and the work W done by the system through the following equation.

$$dU = \delta Q - \delta W$$

Thus, the incremental change in the internal energy of the system is equal to the difference between the incremental amount of heat gained by the system and the incremental amount of work done by the system.

Skill 20.5 Applying the law of conservation of energy

The total energy of the system is defined as the sum of the change in heat energy and the work done by the system. That is to say, the total energy of a system is conserved unless one of the following takes place: heat is added or taken away from the system, or mechanical energy is added to or taken away from the system.

$$dU_{total} = \delta Q - \delta W$$

δQ and δW, are infinitesimal energy amounts rather than true differentials. Also, since δW is defined as the work done by the system, rather than the work done on the system, a minus sign is used. Based on this equation, it can be seen that if the system is isolated thermally and mechanically (i.e., both δQ and δW are equal to zero), then there is no change in the total internal energy of the system, dU_{total}. Internally, then, mechanical energy and energy due to heating are together conserved: energy may be transferred between potential and kinetic energy, or it may be transferred between mechanical and heat energy.

The motion of a moving object can be impeded by the resistance of air, the surfaces of other objects or the viscosity of a liquid. Since energy is conserved, the decrease in mechanical energy in the system is balanced by an increase of heat energy. This heat is energy transferred from motion (kinetic energy) through friction. A swinging pendulum, for example, has alternating kinetic and potential energy, but, over time, this mechanical energy is lost to heat energy by way of friction. If the whole system is thermally and mechanically isolated (i.e., dU = 0), eventually all the mechanical energy of the pendulum will become heat energy. As a side note, it is this phenomenon that, in the context of cosmology, is referred to as "heat death." Some physicists believe that all the "useful" energy of the universe will eventually become heat energy, resulting in a "winding down" of the universe.

In the case of optical systems, the energy contained in incident light can be lost to heat through absorption. On a microscopic level, as described by quantum mechanics, incident photons can be absorbed by a material, causing an excitation of a particular atom or number of atoms. Although these excited atoms may relax to lower energy states through the emission of photons (thus maintaining the energy in the form of light), the relaxation process may also occur through emission of a phonon, or vibrational mode. These phonons are responsible for heat conduction in a material.

For acoustics, sound waves can lose energy to heating as well. As variations in pressure occur, energy in acoustic vibrations can be transferred to heat energy in the material. Again, the total energy of the system must remain constant, and therefore the sum of the mechanical and thermal energies cannot change as long as the system remains closed.

In the case of an ideal gas, which obeys the ideal gas law $PV=nRT$, work done on or by the system result in a transfer of energy available to do work into heat energy. A particular case would be a confined ideal gas; if the volume V of the gas is decreased, meaning work is done on the system, either the pressure P of the gas increases (in which case the gas acquires energy available to do mechanical work) or the heat energy of the gas (as measured by the temperature T) increases. Alternatively, some combination of changes in these properties takes place. If the gas is allowed to cool, venting heat energy out of the gas, then the total energy of the gas decreases, meaning that, in this case, less energy is available for the gas to do work.

Skill 20.6 Analyzing the relationship between entropy and the availability of energy to perform work

To understand the **second law of thermodynamics**, we must first understand the concept of entropy. Entropy is the transformation of energy to a more disordered state and is the measure of how much energy or heat is available for work. The greater the entropy of a system, the less energy is available for work. The simplest statement of the second law of thermodynamics is that the entropy of an isolated system not in equilibrium tends to increase over time. The entropy approaches a maximum value at equilibrium. Below are several common examples in which we see the manifestation of the second law.

- The diffusion of molecules of perfume out of an open bottle
- Even the most carefully designed engine releases some heat and cannot convert all the chemical energy in the fuel into mechanical energy
- A block sliding on a rough surface slows down
- An ice cube sitting on a hot sidewalk melts into a little puddle; we must provide energy to a freezer to facilitate the creation of ice

When discussing the second law, scientists often refer to the "arrow of time". This is to help us conceptualize how the second law forces events to proceed in a certain direction. To understand the direction of the arrow of time, consider some of the examples above; we would never think of them as proceeding in reverse. That is, as time progresses, we would never see a puddle in the hot sun spontaneously freeze into an ice cube or the molecules of perfume dispersed in a room spontaneously re-concentrate themselves in the bottle. The above-mentioned examples are **spontaneous** as well as **irreversible**, both characteristic of increased entropy. Entropy change is zero for a complete cycle in a **reversible process**, a process where infinitesimal quasi-static changes in the absence of dissipative forces can bring a system back to its original state without a net change to the system or its surroundings. All real processes are irreversible. The idea of a reversible process, however, is a useful abstraction that can be a good approximation in some cases.

A quantitative measure of entropy S is given by the statement that the change in entropy of a system that goes from one state to another in an isothermal and reversible process is the amount of heat absorbed in the process divided by the absolute temperature at which the process occurs.

$$\Delta S = \frac{\Delta Q}{T}$$

Stated more generally, the entropy change that occurs in a state change between two equilibrium states A and B via a reversible process is given by

$$\Delta S_{A \to B} = \int_A^B \frac{dQ}{T}$$

Problem: What is the change in entropy of a cube of ice of mass 30g which melts at temperature 0C? The latent heat of fusion of ice is 334 KJ/Kg.

Solution: The amount of heat absorbed by the ice cube =
$30 \times 10^{-3} \times 334 KJ = 10020 J$.

Thus change in entropy = (10020/273)J/K = 36.7 J/K

The second law of thermodynamics may also be stated in the following ways:
1. No machine is 100% efficient.
2. Heat cannot spontaneously pass from a colder to a hotter object.

COMPETENCY 21.0 UNDERSTAND THE BASIC IDEAS OF QUANTUM MECHANICS AND RELATIVITY

Skill 21.1 Explain blackbody radiation and the photoelectric effect

Blackbody Radiation
Until Max Planck presented his theory of energy quantization late in the year 1900, energy was widely believed to consist of a continuous spectrum of values. The failure of this assumption to account for such phenomena as blackbody radiation led Planck to impose on his calculations the notion that energy could only take discrete values; that is, energy is quantized into "energy elements," or quanta.

Planck's theory indicated that the energy in a system is quantized and when the system changes energy states by emitting light, for example, the energy of the light must also have discrete energy values. The consequences of this principle are shown by its ability to account for the behavior of blackbody radiation but a more heuristically beneficial application is the relationship of the quantization of energy to the behavior of the atom.

An atom is composed of a positively charged, heavy nucleus surrounded by an electron "cloud." Since the electrons and the nucleus are oppositely charged, there is an attractive force that would, alone, appear to require that the electrons eventually collide with the nucleus, resulting in a collapse of the atom. From the study of electromagnetism, it is known that when a charged object is accelerated (as an electron would be in an orbit around the nucleus), it radiates energy in the form of electromagnetic waves. Nevertheless, if the possible energy levels of the atom are discrete instead of continuous, then it is feasible that the lowest level does not have sufficient energy to emit a quantum, thus preventing the electron from finally colliding with the nucleus. Therefore, the atom would be stable in spite of the presence of an electron that, since it is still "orbiting," would otherwise be expected to radiate and continually lose energy.

Quantization of energy thus supports atomic theory whereas a purely classical approach involving continuous energy levels would result in a catastrophic failure of the theory. The energy emitted from an atom in the form of light is calculated as the product of the frequency and Planck's constant h. Since h is approximately 6.626×10^{-34} Joule · second, a virtually infinitesimal number, the energies involved at the atomic level are so tiny that they have no apparent impact on the macroscopic world. As a result, quantization effects are not noticeable for macroscopic objects but only become apparent for objects like atoms and subatomic particles.

Planck's theory also leads to the notion that quanta of light, rather than being purely continuous waves as they may appear to be in everyday life, are instead more like "packets" of energy. The concept of these energy packets, or photons, was essential to a more empirically useful understanding of the photoelectric effect, as presented in 1905 by Albert Einstein. Light was then understood as having the characteristics of particles in addition to the characteristics of waves. This concept was also crucial to the development of quantum mechanics in subsequent decades.

The Photoelectric Effect

Einstein used the photon hypothesis to explain the photoelectric effect which is the **emission of electrons from a metal surface when light is incident on it**. When this metal surface is a cathode with the anode held at a higher potential V, the emitted electrons create a current flow in the external circuit. It is observed that current flows only for light of higher frequencies, i.e. electrons are released from the metal only by high frequency light. Also there is a threshold negative potential, the **stopping potential** V_0 below which no current will flow in the circuit.

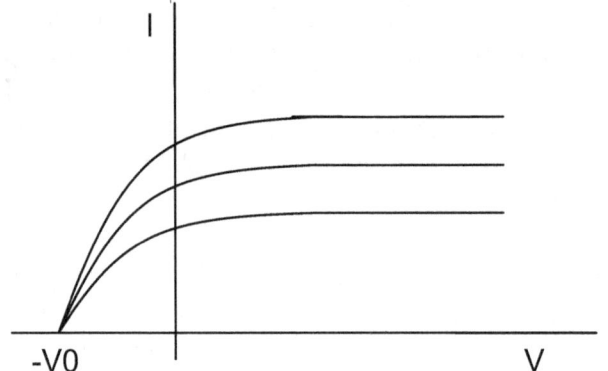

The figure displayed above shows current flow vs. potential for three different intensities of light. It shows that the maximum current flow increases with increasing light intensity but the stopping potential remains the same.

All these observations are counter-intuitive if one considers light to be a wave but may be understood in terms of light particles or photons. According to this interpretation, each photon transfers its energy to a single electron in the metal. Since the energy of a photon depends on its frequency, only a photon of higher frequency can transfer enough energy to an electron to enable it to pass the stopping potential threshold.

When V is negative, only electrons with a kinetic energy greater than $|eV|$ can reach the anode. The maximum kinetic energy of the emitted electrons is given by eV_0. This is expressed by Einstein's photoelectric equation as

$$(\tfrac{1}{2}mv^2)_{max} = eV_0 = hf - \varphi$$

where the **work function** φ is the energy needed to release an electron from the metal and is characteristic of the metal.

Problem: The work function for potassium is 2.20eV. What is the stopping potential for light of wavelength 400nm?

Solution:
$$eV_0 = hf - \varphi = hc/\lambda - \varphi = 4.136 \times 10^{-15} \times 3 \times 10^8 / (400 \times 10^{-9}) - 2.20 = 3.10 - 2.20 = 0.90 \text{eV}$$
Thus stopping potential $V_0 = 0.90$V

Skill 21.2 Describing evidence of the dual nature of light and matter

Scientists have argued for years whether light is a wave or a stream of particles. Actually, light exhibits the behaviors of both waves and particles.

Wave characteristics
Light undergoes reflection, refraction, and diffraction just as any wave would. The image you see in a mirrored surface is the result of the reflection of the light waves off the surface. Light waves follow the "law of reflection," i.e. the angle at which the light wave approaches a flat reflecting surface is equal to the angle at which it leaves the surface. When light crosses the boundary between two different media, its path is bent, or refracted. Diffraction occurs when light encounters an obstacle in its path or passes through an opening. Light diffracts around the sides of an object causing the shadow of the object to appear fuzzy.

Another phenomenon unique to waves is wave interference. This characteristic describes what happens when two waves meet while traveling along the same medium. If light constructively interferes (trough meets trough or crest meets crest), the two light waves reinforce one another to produce a stronger light wave. However, if light destructively interferes (crest meets trough), the two light waves destroy each other and no light wave is produced.

Polarization changes unpolarized light into polarized light. This process can only occur with a transverse wave. An everyday example of polarization is found in polarized sunglasses which reduce glare.

Particle characteristics
Einstein came up with the quantum theory of light that states that light is made up of photons or quanta, discrete particles of electromagnetic radiation. The photons, or individual particles of light, have been shown to have isolated arrival times. A movie was taken of the comet Hyakutake that showed a breakdown of the photons traveling with the comet and scattered throughout the region. Some phenomena such as blackbody radiation and the photoelectric effect can only be explained using the particle nature of light.

Presently, a combination of the two theories, or wave - particle duality, is accepted.

Skill 21.3 Demonstrating a basic understanding of wave functions and the Schrödinger equation

As quantum theory was developed and popularized (primarily by Max Planck and Albert Einstein), chemists and physicists began to consider how it might apply to atomic structure. Niels Bohr put forward a model of the atom in which electrons could only orbit the nucleus in circular orbitals with specific distances from the nucleus, energy levels, and angular momentums. In this model, electrons could only make instantaneous "quantum leaps" between the fixed energy levels of the various orbitals. The Bohr model of the atom was altered slightly by Arnold Sommerfeld in 1916 to reflect the fact that the orbitals were elliptical instead of round.

Though the Bohr model is still thought to be largely correct, it was discovered that electrons do not truly occupy neat, cleanly defined orbitals. Rather, they exist as more of an "electron cloud." The work of Louis de Broglie, Erwin Schrödinger, and Werner Heisenberg showed that an electron can actually be located at any distance from the nucleus. Schrödinger was a Nobel Prize winning Austrian physicist who is best remembered for his contributions to quantum mechanics. Chief among his findings was the Schrödinger equation which describes the space and time-dependence of quantum systems. Schrödinger's work showed that we can find the *probability* that the electrons exists at given energy levels (i.e., in particular orbitals) as mathematically described by wave functions. These probabilities show that the electrons are most frequently organized within the orbitals originally described in the Bohr model.

Skill 21.4 Recognizing models of atomic structure and their relationship to spectroscopy

In the West, the Greek philosophers Democritus and Leucippus first suggested the concept of the atom. They believed that all atoms were made of the same material but that varied sizes and shapes of atoms resulted in the varied properties of different materials. By the 19th century, John Dalton had advanced a theory stating that each element possesses atoms of a unique type. These atoms were also thought to be the smallest pieces of matter which could not be split or destroyed.

Atomic structure began to be better understood when, in 1897, JJ Thompson discovered the electron while working with cathode ray tubes. Thompson realized the negatively charged electrons were subatomic particles and formulated the "plum pudding model" of the atom to explain how the atom could still have a neutral charge overall. In this model, the negatively charged electrons were randomly present and free to move within a soup or cloud of positive charge. Thompson likened this to the dried fruit that is distributed within the English dessert plum pudding though the electrons were free to move in his model.

Ernest Rutherford disproved this model with the discovery of the nucleus in 1909. Rutherford proposed a new "planetary" model of the atom in which electrons orbited around a positively charged nucleus like planets around the sun. Over the next 20 years, protons and neutrons (subnuclear particles) were discovered while additional experiments showed the inadequacy of the planetary model.

As quantum theory was developed and popularized (primarily by Max Planck and Albert Einstein), chemists and physicists began to consider how it might apply to atomic structure. Niels Bohr put forward a model of the atom in which electrons could only orbit the nucleus in circular orbitals with specific distances from the nucleus, energy levels, and angular momentums. In this model, electrons could only make instantaneous "quantum leaps" between the fixed energy levels of the various orbitals. The Bohr model of the atom was altered slightly by Arnold Sommerfeld in 1916 to reflect the fact that the orbitals were elliptical instead of round.

Though the Bohr model is still thought to be largely correct, it was discovered that electrons do not truly occupy neat, cleanly defined orbitals. Rather, they exist as more of an "electron cloud." The work of Louis de Broglie, Erwin Schrödinger, and Werner Heisenberg showed that an electron can actually be located at any distance from the nucleus. However, we can find the *probability* that the electrons exists at given energy levels (i.e., in particular orbitals) and those probabilities will show that the electrons are most frequently organized within the orbitals originally described in the Bohr model.

Atoms and ions of a given element that differ in number of neutrons have a different mass and are called isotopes. In writing nuclear equations, where isotopes of the same element must be distinguished, it is helpful to use a notation where the number of nucleons and protons/electrons is included along with the element symbol. The identity of an element depends on the number of protons in the nucleus of the atom. This value is called the atomic number and it is sometimes written as a subscript before the symbol for the corresponding element. A nucleus with a specified number of protons and neutrons is called a nuclide, and a nuclear particle, either a proton or neutron, may be called a nucleon. The total number of nucleons is called the mass number and may be written as a superscript before the atomic symbol.

$${}^{14}_{6}C$$ represents an atom of carbon with 6 protons and 8 neutrons.

The number of neutrons may be found by subtracting the atomic number from the mass number. For example, uranium-235 has 235−92=143 neutrons because it has 235 nucleons and 92 protons.

An ion is an atom or molecule with a net positive or negative charge (due to an unequal number of protons and electrons) and is represented with a plus or minus sign and the number of excess electrons or protons placed at the top right-hand corner. For example, a positively charged sodium ion with one extra proton may be represented as Na^{+1}.

Quantum #	Radius
$n \to \infty$	$r_\infty \to \infty$
⋮	⋮
$n = 5$	$r_5 = 25a_0$
$n = 4$	$r_4 = 16a_0$
$n = 3$	$r_3 = 9a_0$
$n = 2$	$r_2 = 4a_0$
$n = 1$	$r_1 = a_0$
⊕ (H nucleus)	

An electron may exist at distinct radial distances (r_n) from the nucleus. These distances are proportional to the square of the principal quantum number, n. For a hydrogen atom (shown at left), the proportionality constant is called the Bohr radius ($a_0 = 5.29 \times 10^{-11}$ m). This value is the mean distance of an electron from the nucleus at the ground state of $n = 1$. The distances of other electron shells are found by the formula:

$$r_n = a_0 n^2.$$

As $n \to \infty$, the electron is no longer part of the hydrogen atom. Ionization occurs and the atom become an H^+ ion.

A quantum of energy (ΔE) emitted from or absorbed by an electron transition is directly proportional to the frequency of radiation. The proportionality constant between them is Planck's constant ($h = 6.63 \times 10^{-34}$ J·s):

$$\Delta E = h\nu \quad \text{and} \quad \Delta E = \frac{hc}{\lambda}.$$

The energy of an electron (E_n) is inversely proportional to its radius from the nucleus. For a hydrogen atom, only the principle quantum number determines the energy of an electron by the Rydberg constant ($R_H = 2.18 \times 10^{-18}$ J):

$$E_n = -\frac{R_H}{n^2}.$$

The Rydberg constant is used to determine the energy of a photon emitted or absorbed by an electron transition from one shell to another in the H atom:

$$\Delta E = R_H \left(\frac{1}{n_{initial}^2} - \frac{1}{n_{final}^2} \right).$$

When a photon is absorbed, n_{final} is greater than $n_{initial}$, resulting in positive values corresponding to an endothermic process. Ionization occurs when sufficient energy is added for the atom to lose its electron from the ground state. This corresponds to an electron transition from $n_{initial} = 1$ to $n_{final} \to \infty$. The Rydberg constant is the energy required to ionize one atom of hydrogen. Photon emission causes negative values corresponding to an exothermic process because $n_{initial}$ is greater than n_{final}.

Planck's constant and the speed of light are often used to express the Rydberg constant in units of s^{-1} or length. The formulas below determine the photon frequency or wavelength corresponding to a given electron transition:

$$\nu_{photon} = \left(\frac{R_H}{h}\right)\left|\frac{1}{n_{initial}^2} - \frac{1}{n_{final}^2}\right| \quad \text{and} \quad \lambda_{photon} = \frac{1}{\left(\frac{R_H}{hc}\right)\left|\frac{1}{n_{initial}^2} - \frac{1}{n_{final}^2}\right|}.$$

These formulas relate observed lines in the hydrogen spectrum to individual transitions from one quantum state to another.

A simple optical spectroscope separates visible light into distinct wavelengths by passing the light through a prism or diffraction grating. When electrons in hydrogen gas are excited inside a discharge tube, the emission spectroscope shown below detects photons at four visible wavelengths.

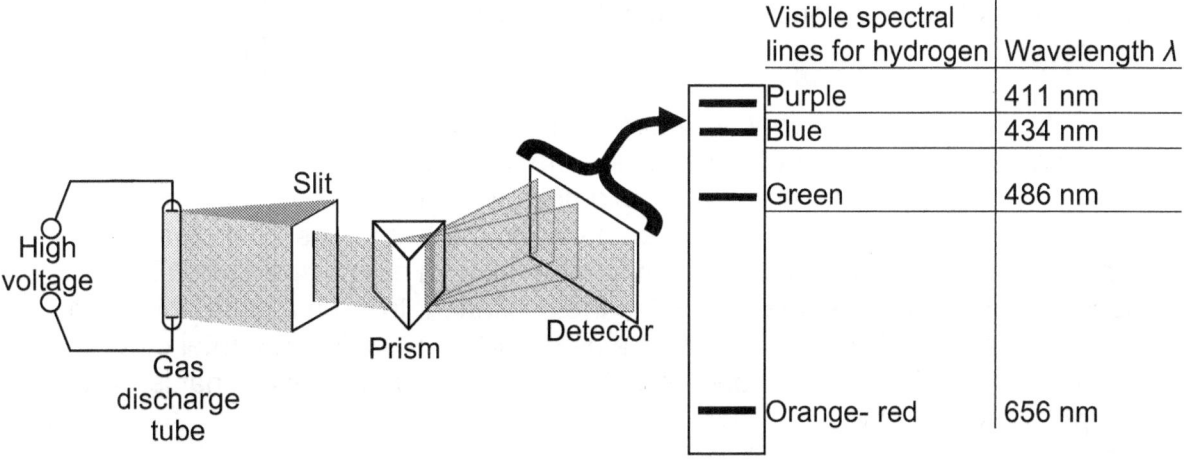

Visible spectral lines for hydrogen	Wavelength λ
Purple	411 nm
Blue	434 nm
Green	486 nm
Orange-red	656 nm

Every line in the hydrogen spectrum corresponds to a transition between electron energy levels. The four spectral lines from hydrogen emission spectroscopy in the visible range correspond to electron transitions from $n = 3, 4, 5,$ and 6 to $n = 2$ as shown in the table below.

Radiation type	Wavelength ×(nm)	Frequency ×(s^{-1})	Energy change ×E (J)	Electron transition $n_{initial} \rightarrow n_{final}$
Ultraviolet	≤397	≥7.55×10^{14}	≤−5.00×10^{-19}	∞→1, ... 2→1 ∞→2, ... 7→2
Purple	411	7.31×10^{14}	−4.84×10^{-19}	6→2
Blue	434	6.90×10^{14}	−4.58×10^{-19}	5→2
Green	486	6.17×10^{14}	−4.09×10^{-19}	4→2
Orange-red	656	4.57×10^{14}	−3.03×10^{-19}	3→2
Infrared and beyond	≥821	≤3.65×10^{14}	≥−2.42×10^{-19}	∞→3, ... 4→3 ∞→4, ... 5→4 ⋮

Most lines in the hydrogen spectrum are not at visible wavelengths. Larger energy transitions produce ultraviolet radiation and smaller energy transitions produce infrared or longer wavelengths of radiation. Transitions between the first three and the first six energy levels of the hydrogen atom are shown in the diagram to the right. The energy transitions producing the four visible spectral lines are colored grey.

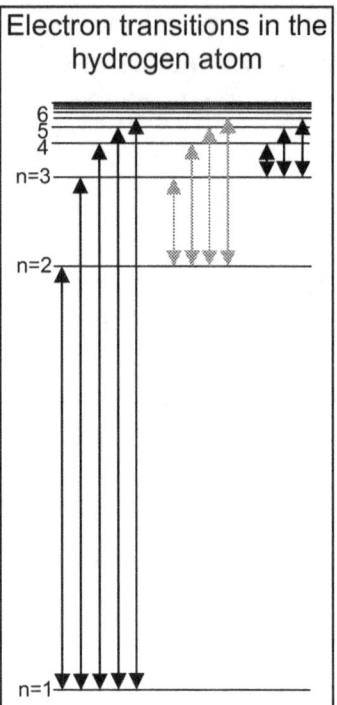

Electron transitions in the hydrogen atom

Skill 21.5   Describing the operation of lasers

The names "laser" and "maser" are acronyms for "light amplification by stimulated emission of radiation" and "microwave amplification by stimulated emission of radiation." Thus, as is evident by their names, these two devices are based on the principle of stimulated emission.

According to quantum theory, an atom in an excited state has a certain probability in a given time frame for relaxing to a lower state through, for example, the emission of a photon of the same energy as the energy difference between the two states. Stimulated emission, however, may take place when the excited atom is perturbed by a passing photon of the same energy as the excitation. In such a case, the atom relaxes to a lower energy by emitting a new photon, resulting in a total of two photons of equal frequency (and, thus, energy) and equal phase. That is to say, the photons are coherent.

In order to produce a significant level of stimulated emission for application in typical lasers and masers, population inversion is required. Population inversion is a situation in which there are more atoms in a particular excited state than in a particular lower-energy state. In order to achieve this condition, the material must be "pumped," which can be performed using electromagnetic fields (light). Pumping to produce population inversion often requires three or more atomic energy levels.

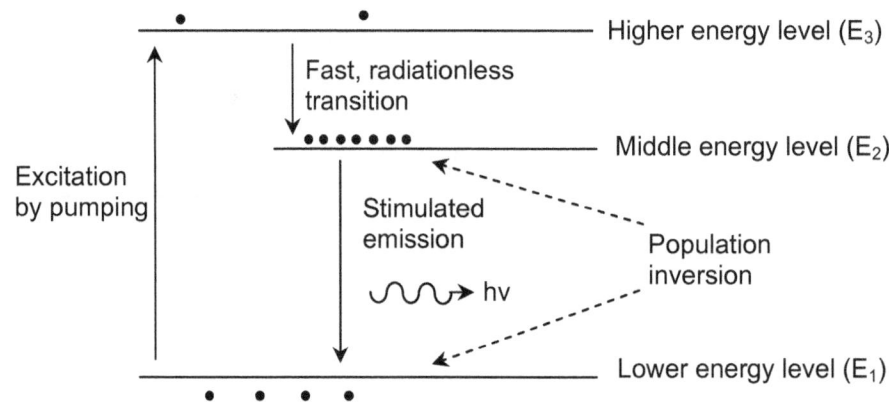

The above energy level diagram is for a so-called three-level laser (or maser). Other lasers may use more levels, depending on the material that is being pumped and the desired frequency output. The transition between the energy levels E_3 and E_2 is noted as being fast and radiationless; that is, almost as soon as the electron(s) of an atom are pumped to E_3, they decay to E_2 without radiating light. The energy difference can be emitted in the form of a phonon (vibration mode in the material, or heat).

The abundance of excited atoms due to population inversion allow for a "chain reaction" to form when photons of the proper frequency (proportional by Planck's constant to the energy difference between the excited state and lower-energy state) are incident. This results in a cascading increase in photon intensity through stimulated emission, thus producing the coherent, high-power light that is used by the laser or maser. This process has its limits, of course, as determined by how quickly the atoms can be pumped in relation to the relaxation processes. A saturation level exists beyond which the rate of stimulated emission cannot be increased. Spontaneously emitted photons can start the cascading process of stimulated emission.

The main components of a laser are a gain medium, an energy source (pump) for the gain medium and a resonant optical cavity with a partial transparency in one mirror. The resonant cavity is tuned to a particular frequency such that the photons of the desired laser frequency are coherent and, largely, isolated within the cavity. A partially transparent mirror on one end of the cavity allows some of the light to exit, thus producing the laser beam.

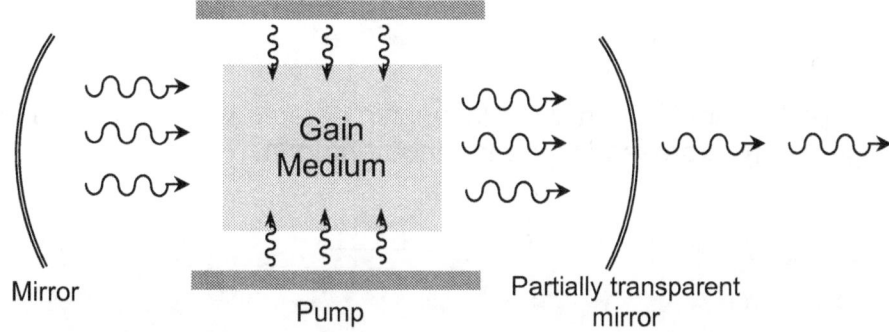

Masers, which operate at lower frequencies than lasers, generally rely on the same principles as lasers, although the types of gain media and resonant cavities may differ.

Skill 21.6 Demonstrating a basic understanding of the theory of special relativity

Postulates of special relativity

Einstein's theory of special relativity built upon the foundation of Galilean relativity and incorporated an analysis of electromagnetics. Galilean relativity is based upon the concept that the laws of physics are invariant with respect to inertial frames of reference. In the context of classical electrodynamics, it is found that Maxwell's equations do not imply any variation in the speed of light with respect to the relative motion of the source and observer. As a result, the second fundamental postulate of special relativity is a statement of the invariance of the speed of light, c, with respect to inertial reference frames.

The postulates of special relativity in brief:
1. Special principle of relativity: The laws of physics are same in all inertial frames of reference.
2. Invariance of c: The speed of light in a vacuum is a universal constant for all observers, regardless of the inertial frame of reference or the relative motion of the source.

Force and acceleration

One of the implications of these postulates is that the relativistic mass m of a particle is dependent upon its velocity v. This relationship between the rest mass m_0 and the relativistic mass is expressed through the Lorentz factor γ.

$$m = \gamma m_0 = \frac{m_0}{\sqrt{1 - \frac{v^2}{c^2}}}$$

This has a particular effect on the relationship of force F and acceleration a, which is generally expressed using the momentum p.

$$F = \frac{\partial p}{\partial t} = \frac{\partial (mv)}{\partial t}$$

In the context of special relativity, where the mass is dependent on the velocity (which is a function of time), the above equation does not simplify to F = ma, as it does in Newtonian mechanics.

$$F = v \frac{\partial m}{\partial t} + ma = m_0 \left(v \frac{\partial \gamma}{\partial t} + \gamma a \right)$$

Since the Lorentz factor increases asymptotically towards infinity as the speed of the particle or object approaches the speed of light, the force required to accelerate the particle also approaches infinity. It would then require infinite energy to accelerate an object to the speed of light, thus making c the "speed limit of the universe." In cases where the speed of the object is significantly less than c, the Lorentz factor is close to unity, thus making Newtonian mechanics an accurate approximation. (This concept is similar to the correspondence principle of quantum mechanics; in this case, relativistic mechanics becomes Newtonian mechanics in the limit as v goes to zero.)

Velocity

As mentioned, the postulates of special relativity seem to imply a universal speed limit. Thus, if this is to be the case in all inertial frames of reference, the observed speed of an object cannot be greater than the speed of light regardless of the velocities of the object and inertial frame of reference as measured in any other inertial frame of reference. As a result, although a particular frame of reference may be moving in one direction at speed w close to light speed and a particular object is moving in the opposite direction at a speed v, likewise near c, the speed of the object as measured from the reference frame *cannot* simply be the sum of v and w. Instead, special relativity uses a different approach to the calculation resulting in a relative velocity given by the following formula.

$$v' = \frac{v + w}{1 + vw/c^2}$$

Conservation of energy

In classical mechanics the behavior of particles or objects can be determined through the application of the conservation of energy and the conservation of momentum. Although, in this case, these two are seemingly independent concepts, in the case of relativistic mechanics they are shown to be interdependent. This interdependence results from the relationship of mass and energy that is derived from the application of the postulates of special relativity.

$$E^2 = (mc^2)^2 + (pc)^2$$

Thus, mass and energy (and, therefore, energy and momentum) are interdependent characteristics of the particle or object. The conservation laws for classical momentum and energy are then joined into a more general expression of the conservation of energy.

COMPETENCY 22.0 UNDERSTAND THE BASIC IDEAS OF NUCLEAR PHYSICS

Skill 22.1 Recognizing models of the nucleus

Though much is known, scientists do not presently have a complete understanding of the structure and properties of the nucleus. It is known that the nucleus contains protons and neutrons and, thus, has a net positive charge. These particles are held together by nuclear force. Electrons and protons themselves are composed of quarks, which are held together by strong interactions. Different elements have nuclei of different sizes with diameters ranging from $1.6 * 10^{-15}$ m to $15 * 10^{-15}$ m. Regardless of the element, however, the nucleus is extremely dense. That is, the nucleus is a very tiny fraction of an atom's volume, but contains almost all its mass.

Many theories have been put forward to help us explain and understand certain observations of atomic nuclei and they are interesting both scientifically and historically. A few of the more prominent theories are listed below:

Liquid drop model

Proposed by George Gamow, this model conceptualized the nucleus as a spherical drop of incompressible fluid. The hypothesized fluid is made of protons and neutrons. This is among the most simple of the nuclear models and leaves many observations unexplained. It does, however, predict the shape of most nuclei and agrees with observations of energy requirements for nuclear fission.

Shell model

The shell model is one of the more rigorous nuclear models and was the result of independent work by Eugene Wigner, Maria Mayer, and J. Hans Jensen. The 1963 Nobel Prize in Physics was awarded to these scientists for work on this model. The shell model posits that the nucleus has shells similar to the electron shells of the atom. These shells are filled with protons and neutrons, with each type of nucleon having its own set of shells. Just as a filled electron shell results in greater stability of an atom, a filled nuclear shell has higher stability than an unfilled one. This theory helps explain why there are certain numbers of nucleons which are more tightly bound; these are known as *magic numbers* (2, 8, 20, 28, 50, 82, and 126).

Moon nuclear model

This model was proposed in 1986 by Robert Moon. It suggests that protons are located at the vertices of geometric solids. These solids are then nested inside one another. This essentially creates energy shells, as suggested by the shell model, but these shells are of varied shape. This model, then, suggests that nuclei have shapes other than the classically assumed spherical one. Like the shell model, the Moon nuclear model explains the phenomenon of magic numbers, but it does not agree well with certain current spectroscopy data. This model continues to be revised, however, and may eventually be improved to the point of explaining that data.

Interacting boson model

Suggested by Iachello and Arima in 1974, this is another more recent model that builds upon the concepts of the shell model. This model, however, attempts to simplify the shell model somewhat. Its basic assumption is that nucleons pair up within the nucleus. These pairs then behave like bosons (force-carrying particles, rather than mass-carrying particles). At present, the theory is useful for determining the nuclear moment and the energy levels of the nucleus. Like the Moon nuclear model, this theory is still rapidly evolving to better explain observed phenomena.

Skill 22.2 Describing properties of nuclei and their applications

Information about the structure of the nucleus can be found in **Skill 22.1.** Scientists are still learning about the structure of the nucleus and the phenomena that take place therein. However, there are a few properties that are fairly well understood at present and have been exploited for technological purposes.

Nuclear spin

Nuclei with odd numbers of protons or neutrons have intrinsic magnetic moments and angular momentums. This makes them sensitive to magnetic fields. Nuclear magnetic resonance technology (NMR) employs this phenomenon. NMR uses a powerful external magnetic field to align the nuclei and then perturbs this alignment. NMR spectroscopy is an extremely powerful tool that is used to determine physical, chemical, electronic and structural information about molecules. Additionally, magnetic resonance imaging (MRI) is built upon this same technology and allows imaging of living organisms, thus having a host of applications in practice of medicine and biological research.

Nuclear decay

Certain nuclei have too few or too many neutrons are thus unstable. This unstable condition ultimately results in decay (loss of mass and/or energy) from the nucleus. How quickly this happens depends on how unstable a given nucleus is. Please see **Skill 22.3** for more information on radioactivity. A variety of technologies have exploited radioactive isotopes for technologies ranging from medical imaging to power generation to dating of geological samples. Also see **Skill 22.5** for more information on these applications.

Skill 22.3   Relating nuclear structure and forces to radioactivity

Nuclear reactions can consist of simple radioactive decay, fission, fusion, and other nuclear processes. In all cases, both **mass-energy and charge must be conserved**. Below we take a closer look at how this is so in the case of alpha and beta decay.

$$^{238}U \rightarrow {}^{234}Th + \alpha$$

The isotope of uranium in the above reaction weighs 238 Daltons. Because it is uranium, it has 92 protons, meaning it must have 146 neutrons. When it undergoes alpha decay it loses 2 protons and 2 neutrons. The alpha particle weighs 4 Daltons and the nucleus that has undergone decay weighs 234 Daltons. Thus mass is conserved. The decayed nucleus will have a charge reduced by that of 2 protons following the decay. However, the emitted alpha particle carries the additional charge due to its 2 protons. Thus both charge and mass are conserved over all.

$$^{137}_{55}Cs \rightarrow {}^{137}_{56}Ba + e^- + \overline{v}_e$$

This isotope of caesium weighs 137 Daltons and, like all caesium isotopes, it has 55 protons. When it undergoes beta minus decay, a neutron is converted to a proton and an electron and an anti-neutrino are lost. The total mass-energy of the system is conserved since the difference in mass between a neutron and an electron plus proton is balanced by the energy of the emitted electron and the anti-neutrino. In beta minus decay a neutron, with no charge, is split into a positively charged proton and a negatively charged electron. Thus the conservation of charge is satisfied. The electron is emitted, while the proton remains in the nucleus. With one extra proton, the nucleus is now a barium isotope.

$$^{22}_{11}Na \rightarrow {}^{22}_{10}Ne + e^+ + v_e$$

In order to conserve mass-energy of the system, beta plus decay cannot occur in isolation but only in a nucleus since the mass of a neutron is greater than the mass of a proton plus electron. The difference in binding energy of the mother and daughter nucleus provides the additional energy needed for the reaction to go through. Charge is conserved when a positively charged proton is converted into a positively charged positron. With one fewer proton, the decayed nucleus is now a neon isotope weighing 22 Daltons.

Problem: Uranium 235 is used as a nuclear fuel in a chain reaction. The reaction is initiated by a single neutron and produces barium 141, an unknown isotope, and 3 neutrons that can go on to propagate the chain reaction. Determine the unknown isotope. Assume that the kinetic energies and the energy released in the reaction is negligible compared to the masses of the isotopes produced.

Solution: We can begin by writing out the reaction, leaving open the unknown isotope X.

$$^{235}_{92}U + ^{1}_{0}n \rightarrow ^{141}_{56}Ba + X + 3^{1}_{0}n$$

We begin with the charge balance; since the neutron has no charge, the unknown isotope must have 36 protons. Consulting a periodic table, we see that this will mean it is a Krypton isotope. Now we can balance the mass. Since the original nucleus weighed 235 Daltons and one neutron was added to it, the total mass of the resultant nuclei must be 236. So, we can simple subtract the weight of the barium isotope and the 3 new neutrons to find the unknown weight:

$$236-141-3=92$$

Thus our unknown isotope is krypton 92, making the balanced equation:

$$^{235}_{92}U + ^{1}_{0}n \rightarrow ^{141}_{56}Ba + ^{92}_{36}Kr + 3^{1}_{0}n$$

TEACHER CERTIFICATION STUDY GUIDE

Skill 22.4 Solving problems involving half-life

While the radioactive decay of an individual atom is impossible to predict, a mass of radioactive material will decay at a specific rate. Radioactive isotopes exhibit exponential decay and we can express this decay in a useful equation as follows:

$$A = A_0 e^{kt}$$

Where A is the amount of radioactive material remaining after time t, A_0 is the original amount of radioactive material, t is the elapsed time, and k is the unique activity of the radioactive material. Note that k is unique to each radioactive isotope and it specifies how quickly the material decays. Sometimes it is convenient to express the rate of decay as half-life. A half-life is the time needed for half a given mass of radioactive material to decay. Thus, after one half-life, 50% of an original mass will have decayed, after two half lives, 75% will have decayed and so on.

Let's examine a sample problem related to radioactive decay.

Problem: Radiocarbon dating has been used extensively to determine the age of fossilized organic remains. It is based on the fact that while most of the carbon atoms in living things is ^{12}C, a small percentage is ^{14}C. Since ^{14}C is a radioactive isotope, it is lost from a fossilized specimen at a specific rate following the death of an organism. The original and current mass of ^{14}C can be inferred from the relative amount of ^{12}C. So, if the half-life of ^{14}C is 5730 years and a specimen that originally contained 1.28 mg of ^{14}C now contains 0.10 mg, how old is the specimen?

In certain problems, we may be simply provided with the activity, k, but in this problem we must use the information given about half-life to solve for k.

Since we know that after one half-life, 50% of the material remains radioactive, we can plug into the governing equation above:

$$A = A_0 e^{kt}$$

$$0.5\, A_0 = A_0 e^{5730k}$$

$$k = (\ln(0.5))/5730 = -0.0001209$$

PHYSICS

Having determined k, we can use this same equation again to determine how old the specimen described above must be:

$$A = A_0 e^{kt}$$

$$0.10 = 1.28 e^{-0.0001209t}$$

$$t = \frac{\ln\left(\frac{0.10}{1.28}\right)}{-0.0001209} = 21087$$

Thus, the specimen is 21,087 years old.

Note that this same equation can be used to calculate the half-life of an isotope if information regarding the decay after a given number of years were provided.

A radioactive isotope often decays into another element that is radioactive as well and continues to decay into a third element. The process continues until a stable element is reached. This chain of disintegration is known as a **decay chain** or a **nuclear disintegration series**. For example, Uranium 238 decays into Radium 226 which decays further into Radon 222.

Skill 22.5 Differentiating between fission and fusion reactions and their applications

Nuclear fission involves the splitting of an atomic nucleus with a subsequent emission of energy in the form of photons with a specific frequency or other particles with a specific kinetic energy. As with combustion, the energy released from a fission reaction can be used to power turbines and to generate electricity. About 20% of the electrical power generated in the United States comes from nuclear reactors.

If controlled, the neutron flux resulting from spontaneously decaying radioactive materials can be used to instigate further decay. With Uranium (^{235}U, specifically), for example, a neutron from another decay can cause the atom to split, thus releasing several neutrons in addition to energy. These neutrons can also cause further decay, resulting in a **chain reaction**. Nuclear weapons involve an uncontrolled chain reaction that yields a tremendously powerful explosion; nuclear power generators take advantage of the same effect, but in a controlled manner.

In order to produce a chain reaction, rods filled with **nuclear fuel pellets** (such as ^{235}U) are used inside the reactor core. Since fission requires relatively slow-moving neutrons, a so-called **moderator** is needed to slow down the fast neutrons produced from fission elsewhere in the core, thus making the chain reaction possible. The moderator can be in the form of water (standard water or deuterium-based "heavy water"), which can also serve as a coolant. While these steps are necessary to sustain a nuclear reaction, it is also equally important to control the reaction. This is accomplished through the use of **control rods** which are filled with a neutron-absorbing material such as boron or silver.

While these materials capture free neutrons, they are not themselves fissile and thus their presence limits the nuclear reaction in the core. The location of the control rods can be adjusted for dynamic control of the reaction. The entire apparatus of fuel rods, control rods and moderator (and coolant) is contained inside a pressure vessel.

The heat produced from the controlled nuclear reaction can be used in the same manner as heat from combustion to produce steam and turn a turbine, thus generating electricity. The hyperboloid towers typically seen at nuclear power plants serve the purpose of cooling.

While nuclear power generation does not produce the same kind of carbon-based emissions as combustion of oil or coal, for example, there are other equally (if not more so) undesirable by-products. Waste products from the nuclear reaction are, themselves, often highly radioactive, posing an environmental threat. Ionizing radiation from such waste can cause tremendous damage to living organisms. Since there is no apparent way to easily neutralize these by-products, they must be stored securely to prevent contamination of the environment. This, of course, requires a storage facility and adequate isolation. Security and structural integrity of these facilities must be continually monitored.

Another issue revolving around the use of nuclear power is the ability to use a nuclear reactor to produce weapons-grade materials for use in explosive nuclear devices. This poses a tremendous threat since nuclear weapons are able to cause untold destruction as seen at the close of the Second World War in Japan. Furthermore, in the context of international terrorism, nuclear power plants in non-aggressive nations can still be a threat if attacked. Thus, the plants themselves, although they may be operating for only peaceful purposes, could be damaged in such a way that nuclear materials are released into the environment, or, in a worst-case scenario, a runaway nuclear reaction is initiated. Even otherwise benign human error can lead to catastrophic consequences. Thus, in light of these considerations, a debate continues concerning nuclear power as the dangers and the benefits are weighed.

Fusion and fission are two complementary reactions with fusion being the preferred reaction for elements lighter than iron (atomic number 26) or nickel (atomic number 28) and fission being the preferred reaction for elements heavier than iron or nickel. In the case of fusion, energy is released for the lighter elements but is absorbed for the heavier elements. Fission operates in the reverse manner.

Fusion reactions and binding energy (mass defect)
The energy released during a fusion reaction is the binding energy which corresponds to a conversion of some of the mass in the constituent particles into emitted energy (mass defect). The mass defect allows the fused nucleus to exist at a lower energy level, thus making it more stable. Although the energy required to fuse two light nuclei is considerable, the energy released by the reaction is even greater and, under the right circumstances, allows for a self-sustaining (chain) reaction. Such a self-sustaining reaction requires that the energy needed to bring two nuclei to within sufficient proximity for fusion to take place is less than the energy liberated when the nuclei are fused. This initial required energy is necessary to overcome the electrostatic repulsion between the positively charged nuclei.

Fusion is believed to be the primary source of solar energy. In a more terrestrial context, fusion has been exploited for the purpose of creating the so-called hydrogen bomb, which involves an uncontrolled chain reaction that releases an immense amount of destructive energy. In the context of the Sun, fusion provides the energy that is a major factor in maintaining the habitability of the Earth.

Stellar fusion
Since fusion takes place in stars, it is a process that is central to the study of astronomy. Two important reactions involved in stellar fusion are the so-called proton-proton (PP) chain and carbon-nitrogen-oxygen (CNO) cycle. The PP chain involves a process whereby protons are converted into heavier helium nuclei with the concomitant release of energy. It is the primary engine of energy production in stars similar to the Sun. The CNO cycle involves interconversion of carbon, nitrogen and oxygen nuclei and is the driving process in larger stars. Although the PP fusion process takes place through a number of stages, the basic components and products of the reaction are described below, where protons (^{1}H) combine to form a helium nucleus (^{4}He), two neutrinos (ν), two positrons (e^+) and energy in the form of photons.

$$4\,^1H \rightarrow\,^4He + 2\nu + 2e^+ + \text{Energy}$$

Skill 22.6 Calculating energy transformations in nuclear reactions

Independently of one another, mass and energy are not necessarily conserved, as one may be converted to another in certain instances. Together, however, as mass-energy, they are indeed conserved. In most macroscopic circumstances, mass and energy are each conserved individually. On a microscopic scale, however, especially in the realm of particle physics, interconversion of mass and energy can take place to a significant extent. In these cases, the total conservation of mass-energy in a closed system is the most general conservation law.

Binding energy and nuclear mass defect
The binding energy and nuclear mass defect are two complementary concepts associated with the interaction of fundamental particles. The protons that form the nucleus of an atom, for example, are positively charged and, thus, have a mutual electrostatic repulsion. As a result, the energy of a system of closely packed protons is high. Nevertheless, the atomic nucleus is still quite stable. This stability results from a release of energy upon formation of the nucleus, thus lowering the total energy of the system. The released energy is called the binding energy. Conversely, for the reverse process, the binding energy is the energy required to split the nucleus into its component particles, and is often described generally as an energy per nucleon.

The source of the released binding energy during a fusion of protons (or protons and neutrons) into a nucleus is a portion of the mass of the system. According to special relativity, the mass of an object (whether rest or relativistic mass) has an equivalent energy ($E = mc^2$ in the case of rest mass). In the context of nuclear physics, protons that are fused into a nucleus are able to release energy at the expense of a portion of their mass. This nuclear mass defect is the emission of a certain portion of the mass of the constituent particles, in the form of energy, to lower the total energy of the whole (e.g., the nucleus). Quantum chromodynamics (QCD) provides a more fundamental explanation of this phenomenon through the strong nuclear force, or color force. QCD details the interactions of the quarks that compose the substructure of protons and neutrons by way of the color charge, which is a property of quarks, and gluons, which are the mediating boson for the strong force.

Conservation of mass-energy
In light of these basic concepts surrounding the conservation of mass-energy, the nuclear mass defect and binding energy can be calculated using the relativistic relationship of mass and energy. The calculation can be performed by noting the difference in mass between the constituent particles and the final product of a reaction.

Problem: What is the binding energy of a deuteron (a nucleus composed of a proton and a neutron)?

Solution: The binding energy is the equivalent energy resulting from the difference in mass between the deuteron and the constituent particles.

Proton mass = 1.6726×10^{-27} kg
Neutron mass = 1.6749×10^{-27} kg
Deuteron mass = 3.3436×10^{-27} kg

Mass difference: $\Delta m = \left(1.6726 \times 10^{-27} + 1.6749 \times 10^{-27}\right) \text{kg} - 3.3436 \times 10^{-27} \text{kg}$
$\Delta m = 3.9 \times 10^{-30}$ kg

Binding energy: $\Delta m c^2 = 3.9 \times 10^{-30} \text{kg} \cdot \left(2.9979 \times 10^8 \, \text{m}/\text{s}\right)^2 = 3.5051 \times 10^{-13}$ J

Skill 22.7 Demonstrate a basic understanding of the properties of quarks and the standard model of elementary particle physics

There are two types of elementary particles: **fermions** and **bosons**. Bosons have integer spin, while fermions have half integer spin.

While there are many subatomic bosons, the most familiar one is the **photon**. A photon has zero mass and charge; in a vacuum, a photon travels at the speed of light. Photons do not spontaneously decay, but can be emitted or absorbed by atoms via a number of natural processes. In fact, photons compose all forms of light and mediate electromagnetic interactions.

The two basic types of fermions are quarks and leptons.

Quarks: Quarks are found nearly exclusively as components of neutrons and protons. They come in three types of arbitrarily named flavors: up, charm, and top (which have a charge of +2/3) and down, strange, and bottom (which have a charge of –1/3. The flavors have varying mass which must be found via indirect methods.

Leptons: Unlike quarks, leptons do not experience strong nuclear force. There are three flavors of leptons; the muon, the tau, and the electron. Each type of lepton consists of a massive charged particle with the same name as the flavor and a smaller neutral particle. This nearly massless neutral particle is called a **neutrino**. Each lepton has a charge of +1 or –1.

Quarks and leptons were introduced in the previous section. Here we analyze how they combine to form the subatomic particles listed below. In each case, we can compare charge, make-up, mass, location, and mobility of the particle.

Electrons: Electrons are a specific type of lepton that exist outside of the nucleus in positions that are functions of their energy levels. Their arrangement and interactions are key to all chemical bonding and reactions. During such chemical processes, electrons can easily be transferred from one atom to another. The charge of an electron is defined as −1 in atomic units (actual charge is -1.6022×10^{-19} coulomb) and the mass of an electron is $1/1836$ of that of a proton.

Protons: Protons are confined inside the atomic nucleus and have a defined charge of +1 atomic units (1.602×10^{-19} coulomb). The mass of a proton is 1 Dalton (or 1 atomic mass unit). Protons may be lost from the nucleus in certain types of radioactive decay. A proton is composed of 2 up quarks and one down quark.

Neutrons: Like protons, neutrons are confined to the nucleus and only lost during certain types of radioactive decay. Neutrons are uncharged particles with assigned mass of 1 Da (in reality neutrons are slightly heavier than protons). A neutron is composed of 2 down quarks and one up quark.

Heavy particles, such as protons and neutrons, that interact by strong interactions are called **hadrons**. **Baryons** are hadrons that are made up of three quarks and are fermions (due to odd number of quarks). Protons and neutrons as well as lambda, sigma, xi and omega particles are baryons. **Mesons** are hadrons that are made up of two quarks and are bosons (due to even number of quarks).

Sample Test

1. A projectile with a mass of 1.0 kg has a muzzle velocity of 1500.0 m/s when it is fired from a cannon with a mass of 500.0 kg. If the cannon slides on a frictionless track, it will recoil with a velocity of ____ m/s.

   A. 2.4

   B. 3.0

   C. 3.5

   D. 1500

2. The weight of an object on the earth's surface is designated x. When it is two earth's radii from the surface of the earth, its weight will be

   A. $x/4$

   B. $x/9$

   C. $4x$

   D. $16x$

3. If a force of magnitude F gives a mass M an acceleration A, then a force $3F$ would give a mass $3M$ an acceleration

   A. A

   B. $12A$

   C. $A/2$

   D. $6A$

4. A car (mass m_1) is driving at velocity v, when it smashes into an unmoving car (mass m_2), locking bumpers. Both cars move together at the same velocity. The common velocity will be given by

   A. m_1v/m_2

   B. m_2v/m_1

   C. $m_1v/(m_1 + m_2)$

   D. $(m_1 + m_2)v/m_1$

5. When acceleration is plotted versus time, the area under the graph represents

   A. Time

   B. Distance

   C. Velocity

   D. Acceleration

6. An inclined plane is tilted by gradually increasing the angle of elevation θ, until the block will slide down at a constant velocity. The coefficient of friction, μ_k, is given by

    A. cos θ

    B. sin θ

    C. cosecant θ

    D. tangent θ

7. Use the information on heats below to solve this problem. An ice block at 0° Celsius is dropped into 100 g of liquid water at 18° Celsius. When thermal equilibrium is achieved, only liquid water at 0° Celsius is left. What was the mass, in grams, of the original block of ice?

    Given: Heat of fusion of ice = 80 cal/g
    Heat of vaporization of ice = 540 cal/g
    Specific Heat of ice = 0.50 cal/g°C
    Specific Heat of water = 1 cal/g°C

    A. 2.0

    B. 5.0

    C. 10.0

    D. 22.5

8. The combination of overtones produced by a musical instrument is known as its

    A. Timbre

    B. Chromaticity

    C. Resonant Frequency

    D. Flatness

9. A long copper bar has a temperature of 60°C at one end and 0°C at the other. The bar reaches thermal equilibrium (barring outside influences) by the process of heat

    A. Fusion

    B. Convection

    C. Conduction

    D. Microwaving

10. The First Law of Thermodynamics takes the form dU = dW when the conditions are

    A. Isobaric

    B. Isochloremic

    C. Isothermal

    D. Adiabatic

11. Given a vase full of water, with holes punched at various heights. The water squirts out of the holes, achieving different distances before hitting the ground. Which of the following accurately describes the situation?

    A. Water from higher holes goes farther, due to Pascal's Principle.

    B. Water from higher holes goes farther, due to Bernoulli's Principle.

    C. Water from lower holes goes farther, due to Pascal's Principle.

    D. Water from lower holes goes farther, due to Bernoulli's Principle.

12. A stationary sound source produces a wave of frequency F. An observer at position A is moving toward the horn, while an observer at position B is moving away from the horn. Which of the following is true?

    A. $F_A < F < F_B$

    B. $F_B < F < F_A$

    C. $F < F_A < F_B$

    D. $F_B < F_A < F$

13. The electric force in Newtons, on two small objects (each charged to −10 microCoulombs and separated by 2 meters) is

    A. 1.0

    B. 9.81

    C. 31.0

    D. 0.225

14. A 10 ohm resistor and a 50 ohm resistor are connected in parallel. If the current in the 10 ohm resistor is 5 amperes, the current (in amperes) running through the 50 ohm resistor is

    A. 1

    B. 50

    C. 25

    D. 60

15. Fahrenheit and Celsius thermometers have the same temperature reading at

    A. 100 degrees

    B. -40 degrees

    C. Absolute Zero

    D. 40 degrees

16. Which of the following apparatus can be used to measure the wavelength of a sound produced by a tuning fork?

    A. A glass cylinder, some water, and iron filings

    B. A glass cylinder, a meter stick, and some water

    C. A metronome and some ice water

    D. A comb and some tissue

17. When the current flowing through a fixed resistance is doubled, the amount of heat generated is

    A. Quadrupled

    B. Doubled

    C. Multiplied by pi

    D. Halved

18. The current induced in a coil is defined by which of the following laws?

    A. Lenz's Law

    B. Burke's Law

    C. The Law of Spontaneous Combustion

    D. Snell's Law

19. If an object is 20 cm from a convex lens whose focal length is 10 cm, the image is:

    A. Virtual and upright

    B. Real and inverted

    C. Larger than the object

    D. Smaller than the object

20. A cooking thermometer in an oven works because the metals it is composed of have different

    A. Melting points

    B. Heat convection

    C. Magnetic fields

    D. Coefficients of expansion

21. In an experiment where a brass cylinder is transferred from boiling water into a beaker of cold water with a thermometer in it, we are measuring

    A. Fluid viscosity

    B. Heat of fission

    C. Specific heat

    D. Nonspecific heat

22. A temperature change of 40 degrees Celsius is equal to a change in Fahrenheit degrees of

   A. 40

   B. 20

   C. 72

   D. 112

23. The number of calories required to raise the temperature of 40 grams of water at 30°C to steam at 100°C is

   A. 7500

   B. 23,000

   C. 24,400

   D. 30,500

24. The boiling point of water on the Kelvin scale is closest to

   A. 112 K

   B. 212 K

   C. 373 K

   D. 473 K

25. The kinetic energy of an object is _____ proportional to its _____.

   A. Inversely...inertia

   B. Inversely...velocity

   C. Directly...mass

   D. Directly...time

26. A hollow conducting sphere of radius R is charged with a total charge Q. What is the magnitude of the electric field at a distance r (given r<R) from the center of the sphere? (k is the electrostatic constant)

   A. 0

   B. kQ/R^2

   C. $kQ/(R^2 - r^2)$

   D. $kQ/(R - r)^2$

27. A quantum of light energy is called a

   A. Dalton

   B. Photon

   C. Curie

   D. Heat Packet

28. The following statements about sound waves are true *except*

   A. Sound travels faster in liquids than in gases.
   B. Sound waves travel through a vacuum.
   C. Sound travels faster through solids than liquids.
   D. Ultrasound can be reflected by the human body.

29. The greatest number of 100 watt lamps that can be connected in parallel with a 120 volt system without blowing a 5 amp fuse is

   A. 24
   B. 12
   C. 6
   D. 1

30. A monochromatic ray of light passes from air to a thick slab of glass (n = 1.41) at an angle of 45° from the normal. At what angle does it leave the air/glass interface?

   A. 45°
   B. 30°
   C. 15°
   D. 55°

31. The magnitude of a force is

   A. Directly proportional to mass and inversely to acceleration
   B. Inversely proportional to mass and directly to acceleration
   C. Directly proportional to both mass and acceleration
   D. Inversely proportional to both mass and acceleration

32. A semi-conductor allows current to flow

   A. Never
   B. Always
   C. As long as it stays below a maximum temperature
   D. When a minimum voltage is applied

33. One reason to use salt for melting ice on roads in the winter is that

   A. Salt lowers the freezing point of water.

   B. Salt causes a foaming action, which increases traction.

   C. Salt is more readily available than sugar.

   D. Salt increases the conductivity of water.

34. Automobile mirrors that have a sign, "objects are closer than they appear" say so because

   A. The real image of an obstacle, through a converging lens, appears farther away than the object.

   B. The real or virtual image of an obstacle, through a converging mirror, appears farther away than the object.

   C. The real image of an obstacle, through a diverging lens, appears farther away than the object.

   D. The virtual image of an obstacle, through a diverging mirror, appears farther away than the object.

35. A gas maintained at a constant pressure has a specific heat which is greater than its specific heat when maintained at constant volume, because

   A. The Coefficient of Expansion changes.

   B. The gas enlarges as a whole.

   C. Brownian motion causes random activity.

   D. Work is done to expand the gas.

36. Consider the shear modulus of water, and that of mercury. Which of the following is true?

   A. Mercury's shear modulus indicates that it is the only choice of fluid for a thermometer.

   B. The shear modulus of each of these is zero.

   C. The shear modulus of mercury is higher than that of water.

   D. The shear modulus of water is higher than that of mercury.

37. A skateboarder accelerates down a ramp, with constant acceleration of two meters per second squared, from rest. The distance in meters, covered after four seconds, is

   A. 10

   B. 16

   C. 23

   D. 37

38. Which of the following units is not used to measure torque?

   A. slug ft

   B. lb ft

   C. N m

   D. dyne cm

39. In a nuclear pile, the control rods are composed of

   A. Boron

   B. Einsteinium

   C. Isoptocarpine

   D. Phlogiston

40. All of the following use semi-conductor technology, except a(n):

   A. Transistor

   B. Diode

   C. Capacitor

   D. Operational Amplifier

41. Ten grams of a sample of a radioactive material (half-life = 12 days) were stored for 48 days and re-weighed. The new mass of material was

   A. 1.25 g

   B. 2.5 g

   C. 0.83 g

   D. 0.625 g

42. When a radioactive material emits an alpha particle only, its atomic number will

   A. Decrease

   B. Increase

   C. Remain unchanged

   D. Change randomly

43. The sun's energy is produced primarily by

    A. Fission

    B. Explosion

    C. Combustion

    D. Fusion

44. A crew is on-board a spaceship, traveling at 60% of the speed of light with respect to the earth. The crew measures the length of their ship to be 240 meters. When a ground-based crew measures the apparent length of the ship, it equals

    A. 400 m

    B. 300 m

    C. 240 m

    D. 192 m

45. Which of the following pairs of elements are not found to fuse in the centers of stars?

    A. Oxygen and Helium

    B. Carbon and Hydrogen

    C. Beryllium and Helium

    D. Cobalt and Hydrogen

46. A calorie is the amount of heat energy that will

    A. Raise the temperature of one gram of water from 14.5º C to 15.5º C.

    B. Lower the temperature of one gram of water from 16.5º C to 15.5º C

    C. Raise the temperature of one gram of water from 32º F to 33º F

    D. Cause water to boil at two atmospheres of pressure.

47. Bohr's theory of the atom was the first to quantize

    A. Work

    B. Angular Momentum

    C. Torque

    D. Duality

48. A uniform pole weighing 100 grams, that is one meter in length, is supported by a pivot at 40 centimeters from the left end. In order to maintain static position, a 200 gram mass must be placed _____ centimeters from the left end.

    A. 10

    B. 45

    C. 35

    D. 50

49. A classroom demonstration shows a needle floating in a tray of water. This demonstrates the property of

   A. Specific Heat

   B. Surface Tension

   C. Oil-Water Interference

   D. Archimedes' Principle

50. Two neutral isotopes of a chemical element have the same numbers of

   A. Electrons and Neutrons

   B. Electrons and Protons

   C. Protons and Neutrons

   D. Electrons, Neutrons, and Protons

51. A mass is moving at constant speed in a circular path. Choose the true statement below:

   A. Two forces in equilibrium are acting on the mass.

   B. No forces are acting on the mass.

   C. One centripetal force is acting on the mass.

   D. One force tangent to the circle is acting on the mass.

52. A light bulb is connected in series with a rotating coil within a magnetic field. The brightness of the light may be increased by any of the following except:

   A. Rotating the coil more rapidly.

   B. Using more loops in the coil.

   C. Using a different color wire for the coil.

   D. Using a stronger magnetic field.

53. The use of two circuits next to each other, with a change in current in the primary circuit, demonstrates

   A. Mutual current induction

   B. Dielectric constancy

   C. Harmonic resonance

   D. Resistance variation

54. A brick and hammer fall from a ledge at the same time. They would be expected to

   A. Reach the ground at the same time

   B. Accelerate at different rates due to difference in weight

   C. Accelerate at different rates due to difference in potential energy

   D. Accelerate at different rates due to difference in kinetic energy

55. The potential difference across a five Ohm resistor is five Volts. The power used by the resistor, in Watts, is

   A. 1

   B. 5

   C. 10

   D. 20

56. An object two meters tall is speeding toward a plane mirror at 10 m/s. What happens to the image as it nears the surface of the mirror?

   A. It becomes inverted.

   B. The Doppler Effect must be considered.

   C. It remains two meters tall.

   D. It changes from a real image to a virtual image.

57. The highest energy is associated with

   A. UV radiation

   B. Yellow light

   C. Infrared radiation

   D. Gamma radiation

58. The constant of proportionality between the energy and the frequency of electromagnetic radiation is known as the

   A. Rydberg constant

   B. Energy constant

   C. Planck constant

   D. Einstein constant

59. A simple pendulum with a period of one second has its mass doubled. If the length of the string is quadrupled, the new period will be

    A. 1 second

    B. 2 seconds

    C. 3 seconds

    D. 5 seconds

60. A vibrating string's frequency is _____ proportional to the _____.

    A. Directly; Square root of the tension

    B. Inversely; Length of the string

    C. Inversely; Squared length of the string

    D. Inversely; Force of the plectrum

61. When an electron is "orbiting" the nucleus in an atom, it is said to possess an intrinsic spin (spin angular momentum). How many values can this spin have in any given electron?

    A. 1

    B. 2

    C. 3

    D. 8

62. Electrons are

    A. More massive than neutrons

    B. Positively charged

    C. Neutrally charged

    D. Negatively charged

63. Rainbows are created by

    A. Reflection, dispersion, and recombination

    B. Reflection, resistance, and expansion

    C. Reflection, compression, and specific heat

    D. Reflection, refraction, and dispersion

64. In order to switch between two different reference frames in special relativity, we use the _____ transformation.

    A. Galilean

    B. Lorentz

    C. Euclidean

    D. Laplace

65. A baseball is thrown with an initial velocity of 30 m/s at an angle of 45°. Neglecting air resistance, how far away will the ball land?

   A. 92 m

   B. 78 m

   C. 65 m

   D. 46 m

66. If one sound is ten decibels louder than another, the ratio of the intensity of the first to the second is

   A. 20:1

   B. 10:1

   C. 1:1

   D. 1:10

67. A wave has speed 60 m/s and wavelength 30,000 m. What is the frequency of the wave?

   A. 2.0×10^{-3} Hz

   B. 60 Hz

   C. 5.0×10^2 Hz

   D. 1.8×10^6 Hz

68. An electromagnetic wave propagates through a vacuum. Independent of its wavelength, it will move with constant

   A. Acceleration

   B. Velocity

   C. Induction

   D. Sound

69. A wave generator is used to create a succession of waves. The rate of wave generation is one every 0.33 seconds. The period of these waves is

   A. 2.0 seconds

   B. 1.0 seconds

   C. 0.33 seconds

   D. 3.0 seconds

70. In a fission reactor, heavy water

   A. Cools off neutrons to control temperature

   B. Moderates fission reactions

   C. Initiates the reaction chain

   D. Dissolves control rods

71. Heat transfer by electromagnetic waves is termed

   A. Conduction

   B. Convection

   C. Radiation

   D. Phase Change

72. Solids expand when heated because

   A. Molecular motion causes expansion

   B. PV = nRT

   C. Magnetic forces stretch the chemical bonds

   D. All material is effectively fluid

73. Gravitational force at the earth's surface causes

   A. All objects to fall with equal acceleration, ignoring air resistance

   B. Some objects to fall with constant velocity, ignoring air resistance

   C. A kilogram of feathers to float at a given distance above the earth

   D. Aerodynamic objects to accelerate at an increasing rate

74. An office building entry ramp uses the principle of which simple machine?

   A. Lever

   B. Pulley

   C. Wedge

   D. Inclined Plane

75. The velocity of sound is greatest in

   A. Water

   B. Steel

   C. Alcohol

   D. Air

76. All of the following phenomena are considered "refractive effects" except for

   A. The red shift

   B. Total internal reflection

   C. Lens dependent image formation

   D. Snell's Law

77. Static electricity generation occurs by

   A. Telepathy

   B. Friction

   C. Removal of heat

   D. Evaporation

78. The wave phenomenon of polarization applies only to

   A. Longitudinal waves

   B. Transverse waves

   C. Sound

   D. Light

79. A force is given by the vector 5 N x + 3 N y (where x and y are the unit vectors for the x- and y- axes, respectively). This force is applied to move a 10 kg object 5 m, in the x direction. How much work was done?

   A. 250 J

   B. 400 J

   C. 40 J

   D. 25 J

80. A satellite is in a circular orbit above the earth. Which statement is false?

   A. An external force causes the satellite to maintain orbit.

   B. The satellite's inertia causes it to maintain orbit.

   C. The satellite is accelerating toward the earth.

   D. The satellite's velocity and acceleration are not in the same direction.

Answer Key

1. B	17. A	33. A	49. B	65. A
2. B	18. A	34. D	50. B	66. B
3. A	19. B	35. D	51. C	67. A
4. C	20. D	36. B	52. C	68. B
5. C	21. C	37. B	53. A	69. C
6. D	22. C	38. A	54. A	70. B
7. D	23. C	39. A	55. B	71. C
8. A	24. C	40. C	56. C	72. A
9. C	25. C	41. D	57. D	73. A
10. D	26. A	42. A	58. C	74. D
11. D	27. B	43. D	59. B	75. B
12. B	28. B	44. D	60. A	76. A
13. D	29. C	45. D	61. B	77. B
14. A	30. B	46. A	62. D	78. B
15. B	31. C	47. B	63. D	79. D
16. B	32. D	48. D	64. B	80. B

Rationales with Sample Questions

1. A projectile with a mass of 1.0 kg has a muzzle velocity of 1500.0 m/s when it is fired from a cannon with a mass of 500.0 kg. If the cannon slides on a frictionless track, it will recoil with a velocity of ____ m/s.

    A. 2.4

    B. 3.0

    C. 3.5

    D. 1500

Answer:

B. 3.0

To solve this problem, apply Conservation of Momentum to the cannon-projectile system. The system is initially at rest, with total momentum of 0 kg m/s. Since the cannon slides on a frictionless track, we can assume that the net momentum stays the same for the system. Therefore, the momentum forward (of the projectile) must equal the momentum backward (of the cannon). Thus:

$p_{projectile} = p_{cannon}$
$m_{projectile} \, v_{projectile} = m_{cannon} \, v_{cannon}$
(1.0 kg)(1500.0 m/s) = (500.0 kg)(x)
x = 3.0 m/s

Only answer (B) matches these calculations.

2. The weight of an object on the earth's surface is designated *x*. When it is two earth's radii from the surface of the earth, its weight will be

    A. *x*/4

    B. *x*/9

    C. 4*x*

    D. 16*x*

Answer:

B. *x*/9

To solve this problem, apply the universal Law of Gravitation to the object and Earth:

$F_{gravity} = (GM_1M_2)/R^2$

Because the force of gravity varies with the square of the radius between the objects, the force (or weight) on the object will be decreased by the square of the multiplication factor on the radius. Note that the object on Earth's surface is *already* at one radius from Earth's center. Thus, when it is two radii from Earth's surface, it is three radii from Earth's center. R^2 is then nine, so the weight is *x*/9.

Only answer (B) matches these calculations.

TEACHER CERTIFICATION STUDY GUIDE

3. If a force of magnitude F gives a mass M an acceleration A, then a force $3F$ would give a mass $3M$ an acceleration

    A. A

    B. $12A$

    C. $A/2$

    D. $6A$

Answer:

A. A

    To solve this problem, apply Newton's Second Law, which is also implied by the first part of the problem:
Force = (Mass)(Acceleration)
$F = MA$
Then apply the same law to the second case, and isolate the unknown:
$3F = 3M\ x$
$x = (3F)/(3M)$
$x = F/M$
$x = A$ (by substituting from our first equation)
Only answer (A) matches these calculations.

TEACHER CERTIFICATION STUDY GUIDE

4. A car (mass m_1) is driving at velocity v, when it smashes into an unmoving car (mass m_2), locking bumpers. Both cars move together at the same velocity. The common velocity will be given by

    A. m_1v/m_2

    B. m_2v/m_1

    C. $m_1v/(m_1 + m_2)$

    D. $(m_1 + m_2)v/m_1$

Answer:

C. $m_1v/(m_1 + m_2)$

In this problem, there is an inelastic collision, so the best method is to assume that momentum is conserved. (Recall that momentum is equal to the product of mass and velocity.)
Therefore, apply Conservation of Momentum to the two-car system:
Momentum at Start = Momentum at End
(Mom. of Car 1) + (Mom. of Car 2) = (Mom. of 2 Cars Coupled)
$m_1v + 0 = (m_1 + m_2)x$
$x = m_1v/(m_1 + m_2)$
Only answer (C) matches these calculations.

Watch out for the other answers, because errors in algebra could lead to a match with incorrect answer (D), and assumption of an elastic collision could lead to a match with incorrect answer (A).

5. **When acceleration is plotted versus time, the area under the graph represents**

    A. Time

    B. Distance

    C. Velocity

    D. Acceleration

Answer:

C. Velocity

The area under a graph will have units equal to the product of the units of the two axes. (To visualize this, picture a graphed rectangle with its area equal to length times width.)
Therefore, multiply units of acceleration by units of time:
(length/time^2)(time)
This equals length/time, i.e. units of velocity.

6. An inclined plane is tilted by gradually increasing the angle of elevation θ, until the block will slide down at a constant velocity. The coefficient of friction, μ_k, is given by

   A. cos θ

   B. sin θ

   C. cosecant θ

   D. tangent θ

Answer:

D. tangent θ

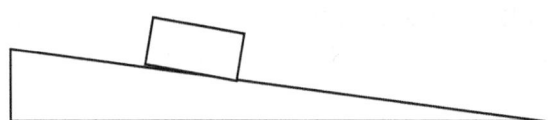

When the block moves, its force upstream (due to friction) must equal its force downstream (due to gravity).

The friction force is given by
$F_f = \mu_k N$
where μ_k is the friction coefficient and N is the normal force.

Using similar triangles, the gravity force is given by
$F_g = mg \sin \theta$
and the normal force is given by
$N = mg \cos \theta$

When the block moves at constant velocity, it must have zero net force, so set equal the force of gravity and the force due to friction:
$F_f = F_g$
$\mu_k mg \cos \theta = mg \sin \theta$
$\mu_k = \tan \theta$

Answer (D) is the only appropriate choice in this case.

7. Use the information on heats below to solve this problem. An ice block at 0° Celsius is dropped into 100 g of liquid water at 18° Celsius. When thermal equilibrium is achieved, only liquid water at 0° Celsius is left. What was the mass, in grams, of the original block of ice?

   Given: Heat of fusion of ice = 80 cal/g
            Heat of vaporization of ice = 540 cal/g
            Specific Heat of ice = 0.50 cal/g°C
            Specific Heat of water = 1 cal/g°C

   A. 2.0

   B. 5.0

   C. 10.0

   D. 22.5

Answer:

D. 22.5

   To solve this problem, apply Conservation of Energy to the ice-water system. Any gain of heat to the melting ice must be balanced by loss of heat in the liquid water. Use the two equations relating temperature, mass, and energy:
   $Q = m\,C\,\Delta T$ (for heat loss/gain from change in temperature)
   $Q = m\,L$ (for heat loss/gain from phase change)
   where Q is heat change; m is mass; C is specific heat; ΔT is change in temperature; L is heat of phase change (in this case, melting, also known as "fusion").

   Then
   $Q_{\text{ice to water}} = Q_{\text{water to ice}}$
   (Note that the ice only melts; it stays at 0° Celsius—otherwise, we would have to include a term for warming the ice as well. Also the information on the heat of vaporization for water is irrelevant to this problem.)
   $m\,L = m\,C\,\Delta T$
   x (80 cal/g) = 100g 1cal/g°C 18°C
   x (80 cal/g) = 1800 cal
   x = 22.5 g

   Only answer (D) matches this result.

TEACHER CERTIFICATION STUDY GUIDE

8. **The combination of overtones produced by a musical instrument is known as its**

    A. Timbre

    B. Chromaticity

    C. Resonant Frequency

    D. Flatness

Answer:

A. Timbre

To answer this question, you must know some basic physics vocabulary. "Timbre" is the combination of tones that make a sound unique, beyond its pitch and volume. (For instance, consider the same note played at the same volume, but by different instruments.) Answer (A) is therefore the only appropriate choice. "Resonant Frequency" is relevant to music, because the resonant frequency of a wave will give the dominant sound tone. "Chromaticity" is an analogous word to "timbre," but it describes color tones. "Flatness" is unrelated, and incorrect.

9. **A long copper bar has a temperature of 60°C at one end and 0°C at the other. The bar reaches thermal equilibrium (barring outside influences) by the process of heat**

    Fusion

    A. Convection

    B. Conduction

    C. Microwaving

Answer:

C. Conduction

To answer this question, recall the different methods of heat transfer. (Note that since the bar is warm at one end and cold at the other, heat must transfer through the bar from warm to cold, until temperature is equalized.) "Convection" is the heat transfer via fluid currents. "Conduction" is the heat transfer via connected solid material. "Fusion" and "Microwaving" are not methods of heat transfer. Therefore the only appropriate answer is (C).

PHYSICS

10. The First Law of Thermodynamics takes the form dU = dW when the conditions are

   A. Isobaric

   B. Isochloremic

   C. Isothermal

   D. Adiabatic

Answer:

D. Adiabatic

To answer this question, recall the First Law of Thermodynamics:
Change in Internal Energy = Work Done + Heat Added
dU = dW + dQ

Thus in the form we are given, dQ has been set to zero, i.e. there is no heat added.
"Adiabatic" refers to a case where there is no heat exchange with surroundings, so answer (D) is the appropriate choice. "Isobaric" means at a constant pressure, "Isothermal" means at a constant temperature, and "Isochloremic" is an imaginary word, as far as I can tell.

It might be tempting to choose "Isothermal," thinking that no heat added would require the same temperature. However, work and internal energy changes can change temperature within the system analyzed, even when no heat is exchanged with the surroundings.

11. Given a vase full of water, with holes punched at various heights. The water squirts out of the holes, achieving different distances before hitting the ground. Which of the following accurately describes the situation?

    A. Water from higher holes goes farther, due to Pascal's Principle.

    B. Water from higher holes goes farther, due to Bernoulli's Principle.

    C. Water from lower holes goes farther, due to Pascal's Principle.

    D. Water from lower holes goes farther, due to Bernoulli's Principle.

Answer:

D. Water from lower holes goes farther, due to Bernoulli's Principle.

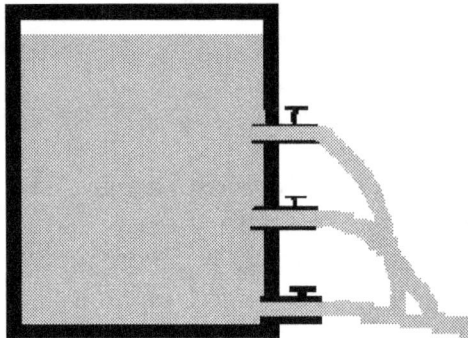

To answer this question, consider the pressure on the water in the vase. The deeper the water, the higher the pressure. Thus, when a hole is punched, the water stream will achieve higher velocity as it equalizes to atmospheric pressure. The lower streams will therefore travel farther before hitting the ground. This eliminates answers (A) and (B). Then recall that Pascal's Principle provides for immediate pressure changes throughout a fluid, while Bernoulli's Principle translates pressure, velocity, and height energy into each other. In this case, the pressure energy is being transformed into velocity energy, and Bernoulli's Principle applies. Therefore, the only appropriate answer is (D).

12. A stationary sound source produces a wave of frequency F. An observer at position A is moving toward the horn, while an observer at position B is moving away from the horn. Which of the following is true?

    A. $F_A < F < F_B$

    B. $F_B < F < F_A$

    C. $F < F_A < F_B$

    D. $F_B < F_A < F$

Answer:

B. $F_B < F < F_A$

To answer this question, recall the Doppler Effect. As a moving observer approaches a sound source, s/he intercepts wave fronts sooner than if s/he were standing still. Therefore, the wave fronts seem to be coming more frequently. Similarly, as an observer moves away from a sound source, the wave fronts take longer to reach him/her. Therefore, the wave fronts seem to be coming less frequently. Because of this effect, the frequency at B will seem lower than the original frequency, and the frequency at A will seem higher than the original frequency. The only answer consistent with this is (B). Note also, that even if you weren't sure of which frequency should be greater/smaller, you could still reason that A and B should have opposite effects, and be able to eliminate answer choices (C) and (D).

13. The electric force in Newtons, on two small objects (each charged to −10 microCoulombs and separated by 2 meters) is

    A. 1.0

    B. 9.81

    C. 31.0

    D. 0.225

Answer:

D. 0.225

To answer this question, use Coulomb's Law, which gives the electric force between two charged particles:
$F = k Q_1 Q_2 / r^2$
Then our unknown is F, and our knowns are:
$k = 9.0 \times 10^9 \text{ Nm}^2/\text{C}^2$
$Q_1 = Q_2 = -10 \times 10^{-6}$ C
$r = 2$ m

Therefore
$F = (9.0 \times 10^9)(-10 \times 10^{-6})(-10 \times 10^{-6})/(2^2)$ N
$F = 0.225$ N

This is compatible only with answer (D).

TEACHER CERTIFICATION STUDY GUIDE

14. A 10 ohm resistor and a 50 ohm resistor are connected in parallel. If the current in the 10 ohm resistor is 5 amperes, the current (in amperes) running through the 50 ohm resistor is

   A. 1

   B. 50

   C. 25

   D. 60

Answer:

A. 1

To answer this question, use Ohm's Law, which relates voltage to current and resistance:
V = IR
where V is voltage; I is current; R is resistance.

We also use the fact that in a parallel circuit, the voltage is the same across the branches.

Because we are given that in one branch, the current is 5 amperes and the resistance is 10 ohms, we deduce that the voltage in this circuit is their product, 50 volts (from V = IR).

We then use V = IR again, this time to find I in the second branch. Because V is 50 volts, and R is 50 ohm, we calculate that I has to be 1 ampere.

This is consistent only with answer (A).

15. Fahrenheit and Celsius thermometers have the same temperature reading at

    A. 100 degrees

    B. - 40 degrees

    C. Absolute Zero

    D. 40 degrees

Answer:

B. - 40 degrees

To answer this question, use the relationship between Fahrenheit and Celsius temperature scales:
F = 9/5 C + 32

Then, in a case where both °F and °C are equal, F = C, so
C = 9/5 C + 32
- 4/5 C = 32
C = - 40

Only answer (B) is consistent with this result.

16. **Which of the following apparatus can be used to measure the wavelength of a sound produced by a tuning fork?**

    A. A glass cylinder, some water, and iron filings

    B. A glass cylinder, a meter stick, and some water

    C. A metronome and some ice water

    D. A comb and some tissue

Answer:

B. A glass cylinder, a meter stick, and some water

To answer this question, recall that a sound will be amplified if it is reflected back to cause positive interference. This is the principle behind musical instruments that use vibrating columns of air to amplify sound (e.g. a pipe organ).
Therefore, presumably a person could put varying amounts of water in the cylinder, and hold the vibrating tuning fork above the cylinder in each case. If the tuning fork sound is amplified when put at the top of the column, then the length of the air space would be an integral multiple of the sound's wavelength. This experiment is consistent with answer (B). Although the experiment would be tedious, none of the other options for materials suggest a better alternative.

17. When the current flowing through a fixed resistance is doubled, the amount of heat generated is

    A. Quadrupled

    B. Doubled

    C. Multiplied by pi

    D. Halved

Answer:

A. Quadrupled

To answer this question, recall that heat generated will occur because of the power of the circuit (power is energy per time). For a circuit with a fixed resistance:
$P = IV$
where P is power; I is current; V is voltage. Then use Ohm's Law:
$V = IR$
where V is voltage; I is current; R is resistance, and substitute:
$P = I^2 R$
and so the doubling of the current I will lead to a quadrupling of the power, and therefore the a quadrupling of the heat.

This is consistent only with answer (A). If you weren't sure of the equations, you could still deduce that with more current, there would be more heat generated, and therefore eliminate answer choice (D) in any case.

18. The current induced in a coil is defined by which of the following laws?

   A. Lenz's Law

   B. Burke's Law

   C. The Law of Spontaneous Combustion

   D. Snell's Law

Answer:

A. Lenz's Law

Lenz's Law states that an induced electromagnetic force always gives rise to a current whose magnetic field opposes the original flux change. There is no relevant "Snell's Law," "Burke's Law," or "Law of Spontaneous Combustion" in electromagnetism. (In fact, only Snell's Law is a real law of these three, and it refers to refracted light.) Therefore, the only appropriate answer is (A).

19. If an object is 20 cm from a convex lens whose focal length is 10 cm, the image is:

   A. Virtual and upright

   B. Real and inverted

   C. Larger than the object

   D. Smaller than the object

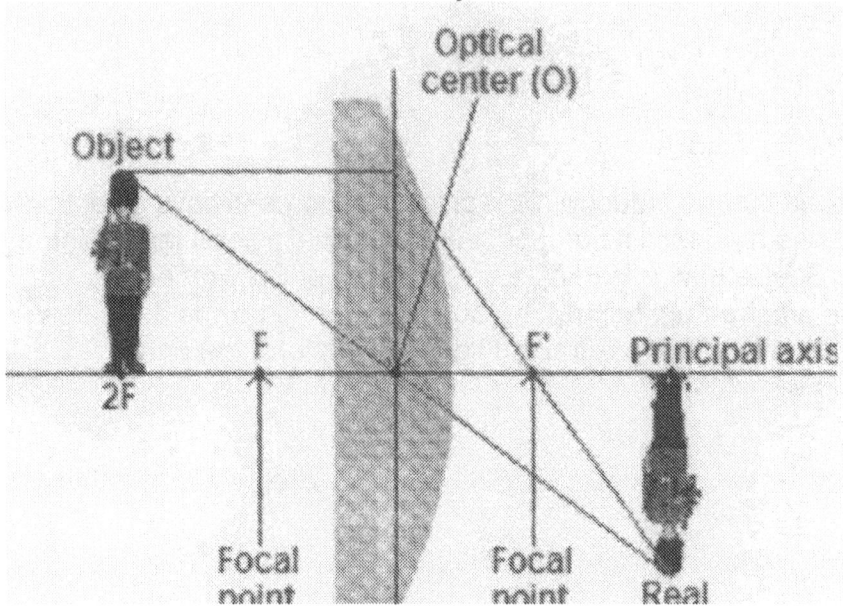

Answer:

B. Real and inverted

To solve this problem, draw a lens diagram with the lens, focal length, and image size.

The ray from the top of the object straight to the lens is focused through the far focus point; the ray from the top of the object through the near focus goes straight through the lens; the ray from the top of the object through the center of the lens continues. These three meet to form the "top" of the image, which is therefore real and inverted. This is consistent only with answer (B).

20. A cooking thermometer in an oven works because the metals it is composed of have different

    A. Melting points

    B. Heat convection

    C. Magnetic fields

    D. Coefficients of expansion

Answer:

D. Coefficients of expansion

A thermometer of the type that can withstand oven temperatures works by having more than one metal strip. These strips expand at different rates with temperature increases, causing the dial to register the new temperature. This is consistent only with answer (D). If you did not know how an oven thermometer works, you could still omit the incorrect answers: It is unlikely that the metals in a thermometer would melt in the oven to display the temperature; the magnetic fields would not be useful information in this context; heat convection applies in fluids, not solids.

21. In an experiment where a brass cylinder is transferred from boiling water into a beaker of cold water with a thermometer in it, we are measuring

    A. Fluid viscosity

    B. Heat of fission

    C. Specific heat

    D. Nonspecific heat

Answer:

C. Specific heat

In this question, we consider an experiment to measure temperature change of water (with the thermometer) as the cylinder cools and the water warms. This information can be used to calculate heat changes, and therefore specific heat. Therefore, (C) is the correct answer. Even if you were unable to deduce that specific heat is being measured, you could eliminate the other answer choices: viscosity cannot be measured with a thermometer; fission takes place at much higher temperatures than this experiment, and under quite different conditions; there is no such thing as "nonspecific heat".

22. A temperature change of 40 degrees Celsius is equal to a change in Fahrenheit degrees of

    A. 40

    B. 20

    C. 72

    D. 112

Answer:

C. 72

To answer this question, recall the equation for Celsius and Fahrenheit:
°F = 9/5 °C + 32

Therefore, whatever temperature difference occurs in °C, it is multiplied by a factor of 9/5 to get the new °F measurement:
new°F = 9/5(old°C + 40) + 32
(whereas old°F = 9/5(old°C) + 32)
Therefore the difference between the old and new temperatures in Fahrenheit is 9/5 of 40, or 72 degrees. This is consistent only with answer (C).

23. The number of calories required to raise the temperature of 40 grams of water at 30°C to steam at 100°C is

   A. 7500

   B. 23,000

   C. 24,400

   D. 30,500

Answer:

C. 24,400

To answer this question, apply the equations for heat transfer due to temperature and phase changes:
Q = mCΔT + mL
where Q is heat; m is mass; C is specific heat; ΔT is temperature change; L is heat of phase change.

In this problem, we are trying to find Q, and we are given:
m = 40 g
C = 1 cal/g°C for water (this should be memorized)
ΔT = 70 °C
L = 540 cal/g for liquid to gas change in water (this should be memorized)

thus Q = (40 g)(1 cal/g°C)(70 °C) + (40 g)(540 cal/g)
Q = 24,400 cal
This is consistent only with answer (C).

24. The boiling point of water on the Kelvin scale is closest to

    A. 112 K

    B. 212 K

    C. 373 K

    D. 473 K

Answer:

C. 373 K

To answer this question, recall that Kelvin temperatures are equal to Celsius temperatures plus 273.15. Since water boils at 100°C under standard conditions, it will boil at 373.15 K. This is consistent only with answer (C).

25. The kinetic energy of an object is _____ proportional to its _____.

    A. Inversely…inertia

    B. Inversely…velocity

    C. Directly…mass

    D. Directly…time

Answer:

C. Directly…mass

To answer this question, recall that kinetic energy is equal to one-half of the product of an object's mass and the square of its velocity:
$KE = \frac{1}{2} m v^2$

Therefore, kinetic energy is directly proportional to mass, and the answer is (C). Note that although kinetic energy is associated with both velocity and momentum (a measure of inertia), it is not *inversely* proportional to either one.

26. A hollow conducting sphere of radius R is charged with a total charge Q. What is the magnitude of the electric field at a distance r (given r<R) from the center of the sphere? (k is the electrostatic constant)

- A. 0
- B. $k\, Q/R^2$
- C. $k\, Q/(R^2 - r^2)$
- D. $k\, Q/(R - r)^2$

Answer:

A. 0

You may be tempted to use the equation for the electric field:
$E = F/Q$ (E = electric field; F = electric force; Q = charge)

and the Coulomb's Law expression for electric force:
$F = k\, Q_1 Q_2 / r^2$ (k = constant; Q_1 and Q_2 = charges; r = distance apart),

which usually would give
$E = k\, Q/(R-r)^2$ in a similar context.

However, this question addresses a special case, i.e. a hollow conductor. Inside a hollow conductor, no electric field exists, because if there were an electric field inside, the conductor's free electrons would be forced to move (by the electric force) until the electric field became zero. Therefore, the only correct answer is (A).

TEACHER CERTIFICATION STUDY GUIDE

27. A quantum of light energy is called a

    A. Dalton

    B. Photon

    C. Curie

    D. Heat Packet

Answer:

B. Photon

The smallest "packet" (quantum) of light energy is a photon. "Heat Packet" does not have any relevant meaning, and while "Dalton" and "Curie" have other meanings, they are not connected to light. Therefore, only (B) is a correct answer.

28. The following statements about sound waves are true *except*

    A. Sound travels faster in liquids than in gases.

    B. Sound waves travel through a vacuum.

    C. Sound travels faster through solids than liquids.

    D. Ultrasound can be reflected by the human body.

Answer:

B. Sound waves travel through a vacuum.

Sound waves require a medium to travel. The sound wave agitates the material, and this occurs fastest in solids, then liquids, then gases. Ultrasound waves are reflected by parts of the body, and this is useful in medical imaging. Therefore, the only correct answer is (B).

29. The greatest number of 100 watt lamps that can be connected in parallel with a 120 volt system without blowing a 5 amp fuse is

    A. 24

    B. 12

    C. 6

    D. 1

Answer:

C. 6

To solve fuse problems, you must add together all the drawn current in the parallel branches, and make sure that it is less than the fuse's amp measure. Because we know that electrical power is equal to the product of current and voltage, we can deduce that:
I = P/V (I = current (amperes); P = power (watts); V = voltage (volts))

Therefore, for each lamp, the current is 100/120 amperes, or 5/6 ampere. The highest possible number of lamps is thus six, because six lamps at 5/6 ampere each adds to 5 amperes; more will blow the fuse.

This is consistent only with answer (C).

30. A monochromatic ray of light passes from air to a thick slab of glass (n = 1.41) at an angle of 45° from the normal. At what angle does it leave the air/glass interface?

   A. 45°

   B. 30°

   C. 15°

   D. 55°

Answer:

B. 30°

To solve this problem use Snell's Law:
$n_1 \sin\theta_1 = n_2 \sin\theta_2$ (where n_1 and n_2 are the indexes of refraction and θ_1 and θ_2 are the angles of incidence and refraction).

Then, since the index of refraction for air is 1.0, we deduce:
$1 \sin 45° = 1.41 \sin x$
$x = \sin^{-1}((1/1.41) \sin 45°)$
$x = 30°$

This is consistent only with answer (B). Also, note that you could eliminate answers (A) and (D) in any case, because the refracted light will have to bend at a smaller angle when entering glass.

31. The magnitude of a force is

   A. Directly proportional to mass and inversely to acceleration

   B. Inversely proportional to mass and directly to acceleration

   C. Directly proportional to both mass and acceleration

   D. Inversely proportional to both mass and acceleration

Answer:

C. Directly proportional to both mass and acceleration

To solve this problem, recall Newton's 2nd Law, i.e. net force is equal to mass times acceleration. Therefore, the only possible answer is (C).

TEACHER CERTIFICATION STUDY GUIDE

32. A semi-conductor allows current to flow

    A. Never

    B. Always

    C. As long as it stays below a maximum temperature

    D. When a minimum voltage is applied

Answer:

D. When a minimum voltage is applied

    To answer this question, recall that semiconductors do not conduct as well as conductors (eliminating answer (B)), but they conduct better than insulators (eliminating answer (A)). Semiconductors can conduct better when the temperature is higher (eliminating answer (C)), and their electrons move most readily under a potential difference. Thus the answer can only be (D).

33. One reason to use salt for melting ice on roads in the winter is that

    A. Salt lowers the freezing point of water.

    B. Salt causes a foaming action, which increases traction.

    C. Salt is more readily available than sugar.

    D. Salt increases the conductivity of water.

Answer:

A. Salt lowers the freezing point of water.

    In answering this question, you may recall that salt is used for road traction because it has large particles to increase traction (analogous to sand). This is true, but salt also has the potential to lower the freezing point of water, thus melting the ice with which it has contact. This is consistent with answer (A). Answer (B) is untrue, and the other two choices, while usually true, are irrelevant in this case.

34. Automobile mirrors that have a sign, "objects are closer than they appear" say so because

    A. The real image of an obstacle, through a converging lens, appears farther away than the object.

    B. The real or virtual image of an obstacle, through a converging mirror, appears farther away than the object.

    C. The real image of an obstacle, through a diverging lens, appears farther away than the object.

    D. The virtual image of an obstacle, through a diverging mirror, appears farther away than the object.

Answer:

D. The virtual image of an obstacle, through a diverging mirror, appears farther away than the object.

To answer this question, first eliminate answer choices (A) and (C), because we have a mirror, not a lens. Then draw ray diagrams for diverging (convex) and converging (concave) mirrors, and note that because the focal point of a diverging mirror is behind the surface, the image is smaller than the object. This creates the illusion that the object is farther away, and therefore (D) is the correct answer.

35. A gas maintained at a constant pressure has a specific heat which is greater than its specific heat when maintained at constant volume, because

    A. The Coefficient of Expansion changes.

    B. The gas enlarges as a whole.

    C. Brownian motion causes random activity.

    D. Work is done to expand the gas.

Answer:

D. Work is done to expand the gas.

To answer this question, recall that the specific heat is a measure of how much energy it takes to raise the temperature of a given mass of gas. Thus, you can reason that when a gas is maintained at constant pressure, some energy is used to expand the volume of the gas, and less is left for temperature changes. In fact, this is the case, and (D) is the correct answer. If you were not able to figure that out, you could still eliminate the other answers, because they are not strictly relevant to a change in specific heat.

36. Consider the shear modulus of water, and that of mercury. Which of the following is true?

    A. Mercury's shear modulus indicates that it is the only choice of fluid for a thermometer.

    B. The shear modulus of each of these is zero.

    C. The shear modulus of mercury is higher than that of water.

    D. The shear modulus of water is higher than that of mercury.

Answer:

B. The shear modulus of each of these is zero.

To answer this question, recall that shear modulus is meaningful only for solids, and that liquids, instead, have a bulk modulus. The only reasonable answer is therefore (B).

37. A skateboarder accelerates down a ramp, with constant acceleration of two meters per second squared, from rest. The distance in meters, covered after four seconds, is

    A. 10

    B. 16

    C. 23

    D. 37

Answer:

B. 16

To answer this question, recall the equation relating constant acceleration to distance and time:
$x = \frac{1}{2} a t^2 + v_0 t + x_0$ where x is position; a is acceleration; t is time; v_0 and x_0 are initial velocity and position (both zero in this case)

thus, to solve for x:
$x = \frac{1}{2} (2 \text{ m/s}^2)(4^2 \text{s}^2) + 0 + 0$
$x = 16$ m

This is consistent only with answer (B).

TEACHER CERTIFICATION STUDY GUIDE

38. Which of the following units is not used to measure torque?

    A. slug ft

    B. lb ft

    C. N m

    D. dyne cm

Answer:

A. slug ft

To answer this question, recall that torque is always calculated by multiplying units of force by units of distance. Therefore, answer (A), which is the product of units of mass and units of distance, must be the choice of incorrect units. Indeed, the other three answers all could measure torque, since they are of the correct form. It is a good idea to review "English Units" before the teacher test, because they are occasionally used in problems.

39. In a nuclear pile, the control rods are composed of

    A. Boron

    B. Einsteinium

    C. Isoptocarpine

    D. Phlogiston

Answer:

A. Boron

Nuclear plants use control rods made of boron or cadmium, to absorb neutrons and maintain "critical" conditions in the reactor. However, if you did not know that, you could still eliminate choice (D), because "phlogiston" is the word for the imaginary element in an ancient structure of earth-air-water-fire.

40. All of the following use semi-conductor technology, except a(n):

    A. Transistor

    B. Diode

    C. Capacitor

    D. Operational Amplifier

Answer:

C. Capacitor

Semi-conductor technology is used in transistors and operational amplifiers, and diodes are the basic unit of semi-conductors. Therefore the only possible answer is (C), and indeed a capacitor does not require semi-conductor technology.

41. Ten grams of a sample of a radioactive material (half-life = 12 days) were stored for 48 days and re-weighed. The new mass of material was

    A. 1.25 g

    B. 2.5 g

    C. 0.83 g

    D. 0.625 g

Answer:

D. 0.625 g

To answer this question, note that 48 days is four half-lives for the material. Thus, the sample will degrade by half four times. At first, there are ten grams, then (after the first half-life) 5 g, then 2.5 g, then 1.25 g, and after the fourth half-life, there remains 0.625 g. You could also do the problem mathematically, by multiplying ten times $(½)^4$, i.e. ½ for each half-life elapsed.

42. When a radioactive material emits an alpha particle only, its atomic number will

    A. Decrease

    B. Increase

    C. Remain unchanged

    D. Change randomly

Answer:

A. Decrease

To answer this question, recall that in alpha decay, a nucleus emits the equivalent of a Helium atom. This includes two protons, so the original material changes its atomic number by a decrease of two.

43. **The sun's energy is produced primarily by**

    A. Fission

    B. Explosion

    C. Combustion

    D. Fusion

Answer:

D. Fusion

To answer this question, recall that in stars (such as the sun), fusion is the main energy-producing occurrence. Fission, explosion, and combustion all release energy in other contexts, but they are not the right answers here.

44. A crew is on-board a spaceship, traveling at 60% of the speed of light with respect to the earth. The crew measures the length of their ship to be 240 meters. When a ground-based crew measures the apparent length of the ship, it equals

    A. 400 m

    B. 300 m

    C. 240 m

    D. 192 m

Answer:

D. 192 m

To answer this question, recall that a moving object's size seems contracted to the stationary observer, according to the equation:
Length Observed = (Actual Length) $(1 - v^2/c^2)^{1/2}$
where v is speed of motion, and c is speed of light.

Therefore, in this case,
Length Observed = $(240 \text{ m}) (1 - 0.6^2)^{1/2}$
Length Observed = (240 m) (0.8) = 192 m

This is consistent only with answer (D). If you were unsure of the equation, you could still reason that because of length contraction (the flip side of time dilation), you must choose an answer with a smaller length, and only (D) fits that description. Note that only the dimension in the direction of travel is contracted. (The length in this case.)

45. Which of the following pairs of elements are not found to fuse in the centers of stars?

    A. Oxygen and Helium

    B. Carbon and Hydrogen

    C. Beryllium and Helium

    D. Cobalt and Hydrogen

Answer:

D. Cobalt and Hydrogen

To answer this question, recall that fusion is possible only when the final product has more binding energy than the reactants. Because binding energy peaks near a mass number of around 56, corresponding to Iron, any heavier elements would be unlikely to fuse in a typical star. (In very massive stars, there may be enough energy to fuse heavier elements.) Of all the listed elements, only Cobalt is heavier than iron, so answer (D) is correct.

46. A calorie is the amount of heat energy that will

    A. Raise the temperature of one gram of water from 14.5° C to 15.5° C.

    B. Lower the temperature of one gram of water from 16.5° C to 15.5° C

    C. Raise the temperature of one gram of water from 32° F to 33° F

    D. Cause water to boil at two atmospheres of pressure.

Answer:

A. Raise the temperature of one gram of water from 14.5° C to 15.5° C.

The definition of a calorie is, "the amount of energy to raise one gram of water by one degree Celsius," and so answer (A) is correct. Do not get confused by the fact that 14.5° C seems like a random number. Also, note that answer (C) tries to confuse you with degrees Fahrenheit, which are irrelevant to this problem.

47. Bohr's theory of the atom was the first to quantize

    A. Work

    B. Angular Momentum

    C. Torque

    D. Duality

Answer:

B. Angular Momentum

Bohr was the first to quantize the angular momentum of electrons, as he combined Rutherford's planet-style model with his knowledge of emerging quantum theory. Recall that he derived a "quantum condition" for the single electron, requiring electrons to exist at specific energy levels.

48. A uniform pole weighing 100 grams, that is one meter in length, is supported by a pivot at 40 centimeters from the left end. In order to maintain static position, a 200 gram mass must be placed _____ \centimeters from the left end.

    A. 10

    B. 45

    C. 35

    D. 50

Answer:

D. 50

In answering this question, do not be tricked into calculating the position of the mass to create balance on the pole's pivot. (This calculation, with equal torques, would lead to incorrect answer (B).) A careful read of the question reveals that we want to "maintain static position" i.e. keep the pole from moving. Because it is already tilted toward the right side, that side (or anywhere to the right of the 45 cm pivot balance answer) is the correct place to put the additional weight without causing the pole to move. Thus, only answer (D) can be correct.

49. **A classroom demonstration shows a needle floating in a tray of water. This demonstrates the property of**

    A. Specific Heat

    B. Surface Tension

    C. Oil-Water Interference

    D. Archimedes' Principle

Answer:

B. Surface Tension

To answer this question, note that the only information given is that the needle (a small object) floats on the water. This occurs because although the needle is denser than the water, the surface tension of the water causes sufficient resistance to support the small needle. Thus the answer can only be (B). Answer (A) is unrelated to objects floating, and while answers (C) and (D) could be related to water experiments, they are not correct in this case. There is no oil in the experiment, and Archimedes' Principle allows the equivalence of displaced volumes, which is not relevant here.

50. **Two neutral isotopes of a chemical element have the same numbers of**

    A. Electrons and Neutrons

    B. Electrons and Protons

    C. Protons and Neutrons

    D. Electrons, Neutrons, and Protons

Answer:

B. Electrons and Protons

To answer this question, recall that isotopes vary in their number of neutrons. (This fact alone eliminates answers (A), (C), and (D).) If you did not recall that fact, note that we are given that the two samples are of the same element, constraining the number of protons to be the same in each case. Then, use the fact that the samples are neutral, so the number of electrons must exactly balance the number of protons in each case. The only correct answer is thus (B).

51. A mass is moving at constant speed in a circular path. Choose the true statement below:

A. Two forces in equilibrium are acting on the mass.

B. No forces are acting on the mass.

C. One centripetal force is acting on the mass.

D. One force tangent to the circle is acting on the mass.

Answer:

C. One centripetal force is acting on the mass.

To answer this question, recall that by Newton's 2nd Law, $F = ma$. In other words, force is mass times acceleration. Furthermore, acceleration is any change in the velocity vector—whether in size or direction. In circular motion, the direction of velocity is constantly changing. Therefore, there must be an unbalanced force on the mass to cause that acceleration. This eliminates answers (A) and (B) as possibilities. Recall then that the mass would ordinarily continue traveling tangent to the circle (by Newton's 1st Law). Therefore, the force must be to cause the turn, i.e. a centripetal force. Thus, the answer can only be (C).

52. A light bulb is connected in series with a rotating coil within a magnetic field. The brightness of the light may be increased by any of the following except:

   A. Rotating the coil more rapidly.

   B. Using more loops in the coil.

   C. Using a different color wire for the coil.

   D. Using a stronger magnetic field.

Answer:

C. Using a different color wire for the coil.

To answer this question, recall that the rotating coil in a magnetic field generates electric current, by Faraday's Law. Faraday's Law states that the amount of emf generated is proportional to the rate of change of magnetic flux through the loop. This increases if the coil is rotated more rapidly (A), if there are more loops (B), or if the magnetic field is stronger (D). Thus, the only answer to this question is (C).

53. The use of two circuits next to each other, with a change in current in the primary circuit, demonstrates

   A. Mutual current induction

   B. Dielectric constancy

   C. Harmonic resonance

   D. Resistance variation

Answer:

A. Mutual current induction

To answer this question, recall that changing current induces a change in magnetic flux, which in turn causes a change in current to oppose that change (Lenz's and Faraday's Laws). Thus, (A) is correct. If you did not remember that, note that harmonic resonance is irrelevant here (eliminating (C)), and there is no change in resistance in the circuits (eliminating (D)).

TEACHER CERTIFICATION STUDY GUIDE

54. **A brick and hammer fall from a ledge at the same time. They would be expected to**

    A. Reach the ground at the same time

    B. Accelerate at different rates due to difference in weight

    C. Accelerate at different rates due to difference in potential energy

    D. Accelerate at different rates due to difference in kinetic energy

Answer:

A. Reach the ground at the same time

This is a classic question about falling in a gravitational field. All objects are acted upon equally by gravity, so they should reach the ground at the same time. (In real life, air resistance can make a difference, but not at small heights for similarly shaped objects.) In any case, weight, potential energy, and kinetic energy do not affect gravitational acceleration. Thus, the only possible answer is (A).

55. The potential difference across a five Ohm resistor is five Volts. The power used by the resistor, in Watts, is

    A. 1

    B. 5

    C. 10

    D. 20

Answer:

B. 5

To answer this question, recall the two relevant equations for potential difference and electric power:
$V = IR$ (where V is voltage; I is current; R is resistance)
$P = IV = I^2R$ (where P is power; I is current; R is resistance)

Thus, first calculate the current from the first equation:
$I = V/R = 1$ Ampere

And then use the second equation:
$P = I^2R = 5$ Watts

This is consistent only with answer (B).

56. An object two meters tall is speeding toward a plane mirror at 10 m/s. What happens to the image as it nears the surface of the mirror?

    A. It becomes inverted.

    B. The Doppler Effect must be considered.

    C. It remains two meters tall.

    D. It changes from a real image to a virtual image.

Answer:

C. It remains two meters tall.

Note that the mirror is a plane mirror, so the image is always a virtual image of the same size as the object. If the mirror were concave, then the image would be inverted until the object came within the focal distance of the mirror. The Doppler Effect is not relevant here. Thus, the only possible answer is (C).

57. The highest energy is associated with

    A. UV radiation

    B. Yellow light

    C. Infrared radiation

    D. Gamma radiation

Answer:

D. Gamma radiation

To answer this question, recall the electromagnetic spectrum. The highest energy (and therefore frequency) rays are those with the lowest wavelength, i.e. gamma rays. (In order of frequency from lowest to highest are: radio, microwave, infrared, red through violet visible light, ultraviolet, X-rays, gamma rays.) Thus, the only possible answer is (D). Note that even if you did not remember the spectrum, you could deduce that gamma radiation is considered dangerous and thus might have the highest energy.

58. The constant of proportionality between the energy and the frequency of electromagnetic radiation is known as the

   A. Rydberg constant

   B. Energy constant

   C. Planck constant

   D. Einstein constant

Answer:

C. Planck constant

Planck estimated his constant to determine the ratio between energy and frequency of radiation. The Rydberg constant is used to find the wavelengths of the visible lines on the hydrogen spectrum.
The other options are not relevant options, and may not actually have physical meaning. Therefore, the only possible answer is (C).

59. A simple pendulum with a period of one second has its mass doubled. If the length of the string is quadrupled, the new period will be

   A. 1 second

   B. 2 seconds

   C. 3 seconds

   D. 5 seconds

Answer:

B. 2 seconds

To answer this question, recall that the period of a pendulum is given by:

$T = 2\pi (L/g)^{1/2}$ where T is period; L is length; g is gravitational acceleration (This is derived from balancing the forces and making small-angle approximation for small angles.)
Note that this equation is independent of mass, so that change is irrelevant. Since the length is quadrupled, and all other quantities on the right side of the equation are constant, the new period will be increased by a factor of two (the square root of four).
This is consistent only with answer (B).

60. A vibrating string's frequency is _____ proportional to the _____.

    A. Directly; Square root of the tension

    B. Inversely; Length of the string

    C. Inversely; Squared length of the string

    D. Inversely; Force of the plectrum

Answer:

A. Directly; Square root of the tension

To answer this question, recall that
$f = (n\, v) / (2\, L)$ where f is frequency; v is velocity; L is length

and

$v = (F_{tension} / (m / L))^{\frac{1}{2}}$ where $F_{tension}$ is tension; m is mass; others as above

so

$f = (n / 2\, L)\, ((F_{tension} / (m / L))^{\frac{1}{2}})$

indicating that frequency is directly proportional to the square root of the tension force. This is consistent only with answer (A). Note that in the final frequency equation, there is an inverse relationship with the square root of the length (after canceling like terms). This is not one of the options, however.

61. When an electron is "orbiting" the nucleus in an atom, it is said to possess an intrinsic spin (spin angular momentum). How many values can this spin have in any given electron?

    A. 1

    B. 2

    C. 3

    D. 8

Answer:

B. 2

To answer this question, recall that electrons fill orbitals in pairs, and the two electrons in any pair have opposite spin from one another. Thus, (B) is correct. Note that answer (D) is trying to mislead you into thinking of the number of valence electrons in an atom.

62. Electrons are

    A. More massive than neutrons

    B. Positively charged

    C. Neutrally charged

    D. Negatively charged

Answer:

D. Negatively charged

Electrons are negatively charged particles that have a tiny mass compared to protons and neutrons. Thus, answer (D) is the only correct alternative.

63. Rainbows are created by

  A. Reflection, dispersion, and recombination

  B. Reflection, resistance, and expansion

  C. Reflection, compression, and specific heat

  D. Reflection, refraction, and dispersion

Answer:

D. Reflection, refraction, and dispersion

To answer this question, recall that rainbows are formed by light that goes through water droplets and is dispersed into its colors. This is consistent with both answers (A) and (D). Then note that refraction is important in bending the differently colored light waves, while recombination is not a relevant concept here. Therefore, the answer is (D).

64. In order to switch between two different reference frames in special relativity, we use the _____ transformation.

  A. Galilean

  B. Lorentz

  C. Euclidean

  D. Laplace

Answer:

B. Lorentz

The Lorentz transformation is the set of equations to scale length and time between inertial reference frames in special relativity, when velocities are close to the speed of light.

The Galilean transformation is a parallel set of equations, used for 'classical' situations when velocities are much slower than the speed of light. Euclidean geometry is useful in physics, but not relevant here. Laplace transforms are a method of solving differential equations by using exponential functions. The correct answer is therefore (B).

TEACHER CERTIFICATION STUDY GUIDE

65. A baseball is thrown with an initial velocity of 30 m/s at an angle of 45°. Neglecting air resistance, how far away will the ball land?

   A. 92 m

   B. 78 m

   C. 65 m

   D. 46 m

Answer:

A. 92 m

To answer this question, recall the equations for projectile motion:
$y = \frac{1}{2} a t^2 + v_{0y} t + y_0$
$x = v_{0x} t + x_0$
where x and y are horizontal and vertical position, respectively; t is time; a is acceleration due to gravity; v_{0x} and v_{0y} are initial horizontal and vertical velocity, respectively; x_0 and y_0 are initial horizontal and vertical position, respectively.
For our case:
x_0 and y_0 can be set to zero
both v_{0x} and v_{0y} are (using trigonometry) = $(\sqrt{2} / 2)$ 30 m/s
$a = -9.81$ m/s^2

We then use the vertical motion equation to find the time aloft (setting y equal to zero to find the solution for t):
$0 = \frac{1}{2} (-9.81 \text{ m/s}^2) t^2 + (\sqrt{2} / 2)$ 30 m/s t
Then solving, we find:
t = 0 s (initial set-up) or t = 4.324 s (time to go up and down)

Using t = 4.324 s in the horizontal motion equation, we find:
$x = ((\sqrt{2} / 2)$ 30 m/s) (4.324 s)
x = 91.71 m

This is consistent only with answer (A).

66. If one sound is ten decibels louder than another, the ratio of the intensity of the first to the second is

    A. 20:1

    B. 10:1

    C. 1:1

    D. 1:10

Answer:

B. 10:1

To answer this question, recall that a decibel is defined as ten times the log of the ratio of sound intensities:
(decibel measure) = $10 \log (I / I_0)$ where I_0 is a reference intensity.

Therefore, in our case,
(decibels of first sound) = (decibels of second sound) + 10
$10 \log (I_1 / I_0) = 10 \log (I_2 / I_0) + 10$
$10 \log I_1 - 10 \log I_0 = 10 \log I_2 - 10 \log I_0 + 10$
$10 \log I_1 - 10 \log I_2 = 10$
$\log (I_1 / I_2) = 1$
$I_1 / I_2 = 10$

This is consistent only with answer (B).
(Be careful not to get the two intensities confused with each other.)

67. A wave has speed 60 m/s and wavelength 30,000 m. What is the frequency of the wave?

    A. 2.0×10^{-3} Hz

    B. 60 Hz

    C. 5.0×10^2 Hz

    D. 1.8×10^6 Hz

Answer:

A. 2.0×10^{-3} Hz

To answer this question, recall that wave speed is equal to the product of wavelength and frequency. Thus:
60 m/s = (30,000 m) (frequency)
frequency = 2.0×10^{-3} Hz

This is consistent only with answer (A).

68. An electromagnetic wave propagates through a vacuum. Independent of its wavelength, it will move with constant

    A. Acceleration

    B. Velocity

    C. Induction

    D. Sound

Answer:

B. Velocity

Electromagnetic waves are considered always to travel at the speed of light, so answer (B) is correct. Answers (C) and (D) can be eliminated in any case, because induction is not relevant here, and sound does not travel in a vacuum.

69. A wave generator is used to create a succession of waves. The rate of wave generation is one every 0.33 seconds. The period of these waves is

    A. 2.0 seconds

    B. 1.0 seconds

    C. 0.33 seconds

    D. 3.0 seconds

Answer:

C. 0.33 seconds

The definition of a period is the length of time between wave crests. Therefore, when waves are generated one per 0.33 seconds, that same time (0.33 seconds) is the period. This is consistent only with answer (C). Do not be trapped into calculating the number of waves per second, which might lead you to choose answer (D).

70. In a fission reactor, heavy water

    A. Cools off neutrons to control temperature

    B. Moderates fission reactions

    C. Initiates the reaction chain

    D. Dissolves control rods

Answer:

B. Moderates fission reactions

In a nuclear reactor, heavy water is made up of oxygen atoms with hydrogen atoms called 'deuterium,' which contain two neutrons each. This allows the water to slow down (moderate) the neutrons, without absorbing many of them. This is consistent only with answer (B).

71. Heat transfer by electromagnetic waves is termed

    A. Conduction

    B. Convection

    C. Radiation

    D. Phase Change

Answer:

C. Radiation

To answer this question, recall the different ways that heat is transferred. Conduction is the transfer of heat through direct physical contact and molecules moving and hitting each other. Convection is the transfer of heat via density differences and flow of fluids. Radiation is the transfer of heat via electromagnetic waves (and can occur in a vacuum). Phase Change causes transfer of heat (though not of temperature) in order for the molecules to take their new phase. This is consistent, therefore, only with answer (C).

72. Solids expand when heated because

    A. Molecular motion causes expansion

    B. PV = nRT

    C. Magnetic forces stretch the chemical bonds

    D. All material is effectively fluid

Answer:

A. Molecular motion causes expansion

When any material is heated, the heat energy becomes energy of motion for the material's molecules. This increased motion causes the material to expand (or sometimes to change phase). Therefore, the answer is (A). Answer (B) is the ideal gas law, which gives a relationship between temperature, pressure, and volume for gases. Answer (C) is a red herring (misleading answer that is untrue). Answer (D) may or may not be true, but it is not the best answer to this question.

73. Gravitational force at the earth's surface causes

   A. All objects to fall with equal acceleration, ignoring air resistance

   B. Some objects to fall with constant velocity, ignoring air resistance

   C. A kilogram of feathers to float at a given distance above the earth

   D. Aerodynamic objects to accelerate at an increasing rate

Answer:

A. All objects to fall with equal acceleration, ignoring air resistance

Gravity acts to cause equal acceleration on all objects, though our atmosphere causes air resistance that slows some objects more than others. This is consistent only with answer (A). Answer (B) is incorrect, because ignoring air resistance leads to the result of constant acceleration, not zero acceleration. Answer (C) is incorrect because all objects (except tiny ones in which random Brownian motion is more significant than gravity) eventually fall due to gravity. Answer (D) is incorrect because it is not related to the constant acceleration due to gravity.

74. An office building entry ramp uses the principle of which simple machine?

   A. Lever

   B. Pulley

   C. Wedge

   D. Inclined Plane

Answer:

D. Inclined Plane

To answer this question, recall the definitions of the various simple machines. A ramp, which trades a longer traversed distance for a shallower slope, is an example of an Inclined Plane, consistent with answer (D). Levers and Pulleys act to change size and/or direction of an input force, which is not relevant here. Wedges apply the same force over a smaller area, increasing pressure—again, not relevant in this case.

TEACHER CERTIFICATION STUDY GUIDE

75. The velocity of sound is greatest in

    A. Water

    B. Steel

    C. Alcohol

    D. Air

Answer:

B. Steel

Sound is a longitudinal wave, which means that it shakes its medium in a way that propagates as sound traveling. The speed of sound depends on both elastic modulus and density, but for a comparison of the above choices, the answer is always that sound travels faster through a solid like steel, than through liquids or gases. Thus, the answer is (B).

76. All of the following phenomena are considered "refractive effects" except for

    A. The red shift

    B. Total internal reflection

    C. Lens dependent image formation

    D. Snell's Law

Answer:

A. The red shift

Refractive effects are phenomena that are related to or caused by refraction. The red shift refers to the Doppler Effect as applied to light when galaxies travel away from observers. Total internal reflection is when light is totally reflected in a substance, with no refracted ray into the substance beyond (e.g. in fiber optic cables). It occurs because of the relative indices of refraction in the materials. Lens dependent image formation refers to making images depending on the properties (including index of refraction) of the lens. Snell's Law provides a mathematical relationship for angles of incidence and refraction. Therefore, the only possible answer is (A).

77. Static electricity generation occurs by

    A. Telepathy

    B. Friction

    C. Removal of heat

    D. Evaporation

Answer:

B. Friction

Static electricity occurs because of friction and electric charge build-up. There is no such thing as telepathy, and neither removal of heat nor evaporation are causes of static electricity. Therefore, the only possible answer is (B).

78. The wave phenomenon of polarization applies only to

    A. Longitudinal waves

    B. Transverse waves

    C. Sound

    D. Light

Answer:

B. Transverse waves

To answer this question, recall that polarization is when waves are screened so that they come out aligned in a certain direction. (To illustrate this, take two pairs of polarizing sunglasses, and note the light differences when rotating one lens over another. When the lenses are polarizing perpendicularly, no light gets through.) This applies only to transverse waves, which have wave parts to align. Light can be polarized, but it is not the only wave that can be. Thus, the correct answer is (B).

79. A force is given by the vector 5 N x + 3 N y (where x and y are the unit vectors for the x- and y- axes, respectively). This force is applied to move a 10 kg object 5 m, in the x direction. How much work was done?

    A. 250 J

    B. 400 J

    C. 40 J

    D. 25 J

Answer:

D. 25 J

To find out how much work was done, note that work counts only the force in the direction of motion. Therefore, the only part of the vector that we use is the 5 N in the x-direction. Note, too, that the mass of the object is not relevant in this problem. We use the work equation:
Work = (Force in direction of motion) (Distance moved)
Work = (5 N) (5 m)
Work = 25 J
This is consistent only with answer (D).

TEACHER CERTIFICATION STUDY GUIDE

80. A satellite is in a circular orbit above the earth. Which statement is false?

    A. An external force causes the satellite to maintain orbit.

    B. The satellite's inertia causes it to maintain orbit.

    C. The satellite is accelerating toward the earth.

    D. The satellite's velocity and acceleration are not in the same direction.

Answer:

B. The satellite's inertia causes it to maintain orbit.
To answer this question, recall that in circular motion, an object's inertia tends to keep it moving straight (tangent to the orbit), so a centripetal force (leading to centripetal acceleration) must be applied. In this case, the centripetal force is gravity due to the earth, which keeps the object in motion. Thus, (A), (C), and (D) are true, and (B) is the only false statement.

XAMonline, INC. 21 Orient Ave. Melrose, MA 02176
Toll Free number 800-509-4128
TO ORDER Fax 781-662-9268 OR www.XAMonline.com

MICHIGAN TEST FOR TEACHER EXAMINATION - MTTC - 2007

PO#	Store/School:
Address 1:

Address 2 (Ship to other):

City, State Zip

Credit card number _____-_____-_____-_____	expiration _____
EMAIL _____
PHONE	FAX

13# ISBN 2007	TITLE	Qty	Retail	Total
978-1-58197-968-8	MTTC Basic Skills 96			
978-1-58197-954-1	MTTC Biology 17			
978-1-58197-955-8	MTTC Chemistry 18			
978-1-58197-957-2	MTTC Earth-Space Science 20			
978-1-58197-966-4	MTTC Elementary Education 83			
978-1-58197-967-1	MTTC Elementary Education 83 Sample Questions			
978-1-58197-950-3	MTTC English 02			
978-1-58197-961-9	MTTC Family and Consumer Sciences 40			
978-1-58197-959-6	MTTC French Sample Test 23			
978-1-58197-965-7	MTTC Guidance Counselor 51			
978-1-58197-964-0	MTTC Humanities& Fine Arts 53, 54			
978-1-58197-972-5	MTTC Integrated Science (Secondary) 94			
978-1-58197-973-2	MTTC Emotionally Impaired 59			
978-1-58197-953-4	MTTC Learning Disabled 63			
978-1-58197-963-3	MTTC Library Media 48			
978-1-58197-958-9	MTTC Mathematics (Secondary) 22			
978-1-58197-962-6	MTTC Physical Education 44			
978-1-58197-665-6	MTTC Physics 19			
978-1-58197-952-7	MTTC Political Science 10			
978-1-58197-951-0	MTTC Reading 05			
978-1-58197-960-2	MTTC Spanish 28			
978-158197-970-1	MTTC Social Studies 84			
	FOR PRODUCT PRICES GO TO WWW.XAMONLINE.COM		**SUBTOTAL**	
			Ship	$8.25
			TOTAL	

www.ingramcontent.com/pod-product-compliance
Lightning Source LLC
Chambersburg PA
CBHW080537300426
44111CB00017B/2769